AF777553

Ecosystem-Based Fisheries Management

Confronting Tradeoffs

Responsible fisheries management is of increasing interest to the scientific community, resource managers, policy makers, stakeholders, and the general public. Focusing solely on managing one species of fish stock at a time has become less of a viable option for many reasons. Incorporating more holistic considerations into fisheries management by addressing the tradeoffs among the range of issues involved, such as ecological principles, legal mandates, and the interests of stakeholders, will hopefully challenge and shift the perception that ecosystem-based fisheries management (EBFM) is unfeasible. Demonstrating that EBFM is, in fact, feasible will have widespread impact, in both US and international waters. Using case studies, examining underlying philosophies, and exploring analytical approaches, this book brings together a range of interdisciplinary topics surrounding EBFM and considers these simultaneously, with the aim of providing tools for successful implementation and of furthering the debate on EBFM, ultimately hoping to foster enhanced living marine resource management.

Jason S. Link is currently a Senior Research Fisheries Biologist at the National Marine Fisheries Service in Woods Hole, USA. He has spent a large part of his career helping to establish the scientific underpinnings for EBFM and has received the Fisheries Society of the British Isles Medal for significant advances in fisheries science.

Ecosystem-Based Fisheries Management

Confronting Tradeoffs

JASON S. LINK
National Marine Fisheries Service, Woods Hole, MA, USA

CAMBRIDGE
UNIVERSITY PRESS

CAMBRIDGE UNIVERSITY PRESS
Cambridge, New York, Melbourne, Madrid, Cape Town, Singapore,
São Paulo, Delhi, Dubai, Tokyo, Mexico City

Cambridge University Press
The Edinburgh Building, Cambridge CB2 8RU, UK

Published in the United States of America by
Cambridge University Press, New York

www.cambridge.org
Information on this title: www.cambridge.org/9780521762984

© J. S. Link 2010

This publication is in copyright. Subject to statutory exception
and to the provisions of relevant collective licensing agreements,
no reproduction of any part may take place without
the written permission of Cambridge University Press.

First published 2010

Printed in the United Kingdom at the University Press, Cambridge

A catalog record for this publication is available from the British Library

Library of Congress Cataloging-in-Publication Data

Link, Jason S.
Ecosystem-based fisheries management : confronting tradeoffs / J.S. Link.
p. cm.
Includes bibliographical references and index.
ISBN 978-0-521-76298-4
1. Fishery management. 2. Marine fishes–Ecology. 3. Fish populations.
I. Title.
SH328.L564 2010
333.95′6–dc22 2010016808

ISBN 978-0-521-76298-4 Hardback

Cambridge University Press has no responsibility for the persistence or
accuracy of URLs for external or third-party internet websites referred to
in this publication, and does not guarantee that any content on such
websites is, or will remain, accurate or appropriate.

Contents

Preface

The subtitle of this book conveys the essence of what ecosystem-based management is all about: confronting tradeoffs.

Here is a question I have been asked, in one form or another, about once a month for the past five years: Will there be enough herring (i.e. a key forage fish species) left for all of the fisheries that target them, the whales that eat them, the piscivorous fish that also eat them, the plankton that are eaten by them to be cropped down, and the jellyfish that compete with them to be kept in check; all while nutrients and water temperatures are notably changing? This one example of a focal species makes the tradeoffs among a suite of issues readily apparent – and this is just one instance among a plethora of such examples.

It has become abundantly clear that we need to explore and address these tradeoffs. Ignoring them (particularly via focusing solely on one stock at a time) has become singularly less of a viable option, so allow me to explain how all these thoughts began to coalesce in my ponderings of these kinds of issues.

Some of the first formal stock assessments I sat in on were quite shocking to me, a scientist primarily trained as an ecologist with a particular interest in predator–prey interactions. Aside from the incredible stakes of the results, which led to palpable animosity from all attending stakeholders, what struck me were the assumptions of natural mortality. This term for mortality, M, was constant – both age and time invariant – and set at an amazingly low level for some stocks that were clearly a significant forage base for many other species. To make a long story short, this issue led to a series of discussions, reading groups, and presentations on what were, from my perspective, some basic ecological principles, and to a poor attempt on my part to contextualize those principles into an arena that was a hybrid of fisheries science and management. Over time these endeavors led to multiple discussions and my publishing two articles (Link 2002b, 2002c) that attempted to note the importance of other factors that can influence fish stock dynamics. My hope was that those publications would be the harbinger of subtle change, leading to the ready uptake of ecological principles. Suffice it to say, this broad range of ecosystem issues has continued to escalate and the uptake of ecological principles in a fisheries context has remained perhaps a little less than escalated.

Throughout the past decade or so I have published several more articles, sat on scores of review panels, been involved in multiple working groups and workshops, and given countless talks around the world on the topics contained in this book. From those endeavors it became apparent that there was no one place that pulled together the various and myriad aspects of ecosystem-based fisheries management (EBFM) to foster its implementation in an operational, pragmatic manner. Clearly, continued discussions and shorter, focal articles would help at some level.

But it also became clear that such efforts would ultimately not achieve the fully desired effect. I kept waiting for various groups or individuals more august than I to pull a comprehensive tome on EBFM together. Although several excellent extant works have been produced that are related to the topic, there has never been anything that spanned the full range of issues, was not overly focused on more specific issues, nor was beyond more than general principles or "platitudinals" on the topic. In short, nothing seemed to capture the need and frustration that I and my global colleagues had expressed; there was nothing to provide a "how-to" manual for doing EBFM. Moreover, many of the stakeholders interested in fisheries management issues that I interact with also noted a need for a clear, concise treatment of the subject.

Thus this work ultimately resulted from my perception of a need to compile these composite ideas formed over many years into one such place. Having created one place to focus the debate and discussion for EBFM (although I certainly have not even remotely treated every topic fully, adequately, or perhaps even much beyond a naïve simplification), this will provide both the scientific discipline and the resource management applications based thereon a tool from which further implementation of EBFM can be expedited and facilitated.

Again, most of the publications, review panels, workshops, and presentations I participated in were part of the professional duties of my job. Therefore, it is hard to separate out and attribute the development and generation of many of the thoughts, concepts, and ideas put forth in this book, as I thought about these topics while at work. Yet I also need to be clear: the thoughts and ideas contained herein are my own and do not necessarily represent those of my employer (even though I give my professional affiliation). Since I firmly believe in living a life of integrity, I note that although at times I used a laptop from my employer to craft this book, I wrote it on my own time – in hotels at odd hours while fighting jet lag; before going into the lab many mornings; while on vacation – in short, anytime I could spare an hour or two. I need to reiterate that this book was written by me as a private individual and does not necessarily reflect the views of my employer, the National Marine Fisheries Service (though I legitimately hope they and similar resource management agencies will eventually adopt many of these views). I also must admit this distinction because although I do not (and mostly did not feel the need to) directly say things such as "the agency messed up on that one" or "that was stupid," I felt I needed the freedom to be able to challenge instances where it was appropriate to question or point out areas that need improvement – particularly as it referenced institutional and philosophical items – from my limited and humble perspective. Having this work be a product of "on-the-clock" work efforts could potentially subvert any integrity in doing it, from both the agency's or an external observer's perspective. And as it had to be done in this way, on my own time, some of my professional friends who have known about this effort for years will note that it has taken longer than I hoped to complete this work. For that I apologize.

On that note, allow me to provide a disclaimer: portions of this book were written in what may be an atypical scientific style. It's not that the material is unscientific or not technically solid; rather the presentation at times can tend towards levity, subtlety, or a conversational style not typical in scientific writings. I do so because the topic can be emotionally charged with high-stakes outcomes, and I find that a relaxed presentation and well-placed humor can diffuse such tense situations. Furthermore, I as the author and you as the reader might as well enjoy the presentation of this material in what could otherwise become a set of quite dry and boring theses. I have attempted always to be respectful of positions, perspectives, and persons, but also have no problem gently pointing out silliness in logic or practice, if appropriate.

Another disclaimer is that my primary professional experience has been in the USA and particularly the north Atlantic. I have attempted to be as inclusive of other examples, perspectives, and situations as I could be. I have traveled to panels abroad and reviewed a wide range of global efforts on the topic, but although familiar with other situations, I do not work there day-in and day-out. The obvious focus here is on marine ecosystems, but I clearly recognize the appropriateness of the application of these concepts to freshwater systems as well. Thus I readily admit and strongly suspect that my biases have shaped how I view the implementation of EBFM (as compared, for example, to spending my time working in a remote, developing nation with a small fishery) and as such, I want to acknowledge my perspective and my potentially limited view, all up front.

This book, like most others, was not written in a vacuum. I admit to drawing heavily on my colleague Steve Hall's 1999 book and the multiple combinations of Simon Jennings' and Michel Kaiser's books/review papers (e.g. 1998) on the topic. My colleague Tim Smith's (1994) history of fishes notably enlightened my view of how fisheries are prosecuted and how fisheries science has been conducted. Another book by Charles (2001) was also quite helpful, filling in many of the gaps in my knowledge of fisheries management systems. I consider my effort here as a natural extension and merger of those prior works, and thank those authors for their concepts.

I also must thank a broad host of professional colleagues who have worked, conversed, or corresponded with me over the years and in so doing have challenged, disagreed with, encouraged and stimulated my thinking and efforts for this book: Tim Smith, Tony Smith, Beth Fulton, Alida Bundy, my favorite doctor bill – Bill Overholtz, Mike Fogarty, Marc Mangel, Keith Sainsbury, Lance Garrison, Lisa Methratta, Hassan Moustahfid, Megan Tyrrell, Brian Smith, Rob Gamble, Janet Nye, Jon Hare, Kevin Friedland, Jon Brodziak, Sarah Gaichas, John Field, Tom Miller, Ed Houde, Beth Babcock, Fabio Pranovi, Simone Libralato, Marta Coll, Tony Pitcher, J J Maguire, Ken Sherman, Michel Kaiser, Bjarte Bogstad, Eva Plagányi-Lloyd, Anne Hollowed, Libby Logerwell, Villy Christensen, Steve Martell, Kerim Aydin, Howard Townsend, Jay O'Reilly, Jack Green, Vince Guida, Joe Vitaliano, Deb Palka, Laurel Col, Robin Griswold, Peter Auster, my good friend the late Steve Edwards, Dan Georgianna, John Walden, Jeremy Collie, Rob LaTour, Mike Ford, Phil Levin, Isaac Kaplan, Chris Harvey, Bern Megrey, Andy Payne, Steve Mackinson, Julia Blanchard, Simon Jennings, Yunne Shin, Philippe Cury, Nancy Shackell, John Pinnegar, Mariano Koen-Alonso, Andy Belgranno, Buck Stockhausen, George Watters, Steve Cadrin, Marie Joelle-Rochet, Verena Trenkel, Jake Rice, Jae Choi, Ken Frank, Andy Rosenberg, Les Kaufman, Steve Murawski, and so many others who have helped in discussions on the topic. I particularly thank Marc Mangel and Les Kaufman who have been extremely encouraging of my efforts to write this book. I also warmly thank my friends and colleagues at the Waquoit Bay National Estuarine Research Reserve who allowed me to hole up in a spare room during my Christmas holiday and a slow point in their calendar to help finish up some of the final steps needed for completing this book. Additionally, I have presented a trial run of this material in graduate courses I have taught at the University of Massachusetts Dartmouth's School for Marine Science and Technology (SMAST), at the Venice International University, and at the University of Venice; to the students in those classes whose insightful and probing questions helped to sharpen my focus on particular points and issues, I particularly extend my thanks.

I also thank the many fisheries scientists who take a different, usually more classical and single-species perspective on the topic. These individuals and institutions may have disagreed with some of the thoughts in this book over the years, but in so doing have forced me to evaluate those concepts more rigorously than I otherwise would have. To all who have made a professional and institutional contribution, I provide a hearty thank you.

I thank my editors, Dominic Lewis, Janice Robertson, Sabine Koch, Lynette Talbot, and Rachel Eley, at Cambridge University Press. Their cheerful dedication and assistance have been extremely helpful in the completion of this project. I thank Lillian Lomba who helped to compile some of the materials contained in this book from a wide variety of my notes, lectures, and reprints. I especially thank Jessie Gunnard who provided notable and very much appreciated assistance in the preliminary compilation, organization, and preliminary editing of this book.

These interactions have all been invaluable to the development of the concepts presented in this book. Yet as much as I acknowledge the contributions of my many professional and editorial colleagues, I take full responsibility for the contents of this book. Any errors herein are mine.

This book, also like many others, was the result of due diligence and patience by the many longsuffering individuals in my life. I especially thank my wife and children for allowing me to work on this during portions of family vacations, weekend afternoons, and at all odd hours when I probably should have been doing other familial duties. I appreciate their patience and understanding. Finally, I acknowledge that I am a man of faith and as such want to thank God for allowing us to enjoy, study, and research the fascinating fishes and creatures of the world's oceans. I also want to thank God for trusting those of us engaged in the topics of this book enough to be stewards of these great oceans and their inhabiting life. May we be found to be faithful stewards of such resources.

My last disclaimer is that if I have omitted something in this book, or have inaccurately represented a situation, or have otherwise missed the point on a particular topic, I apologize now and ask for your indulgence. Given the highly interdisciplinary nature of the topic, I would be surprised if I did *not* make such mistakes. That is, I am not sure if I have gotten the treatment of this topic entirely right and it is with a spirit of humility that I present the material herein. I trust that the value of this book is seen in its interdisciplinariness – not being exhaustive or conclusive on any topic, but rather being representative of a range of topics pulled together and considered simultaneously. I trust that you will view this effort in the vein in which it was intended; if there are such errors, misperceptions, oversimplifications, or miscommunications, let us allow this book to stimulate further dialogue among those interested parties so that we can continue to advance the scientific discipline and management practice of EBFM.

Part I Context

This section provides the historical, theoretical, philosophical, and contextual background for why we would want to consider doing ecosystem-based fisheries management (EBFM). Admittedly, many of these areas are outside my area of expertise, but there have been several delightful summaries, histories, and perspectives written on the topic from which I was able to draw quite heavily. I have written this section to set the stage and provide some background before we begin some of the more technical aspects (read: math!) on the topic of EBFM.

This section starts by noting that what we are doing for fisheries management has not always been sufficient, and highlights the need for more holistic and broader approaches. Following that, the next chapter seeks to define our terms and clarify many of the concepts surrounding this topic.

The third chapter explores many of the underlying philosophies about our stewardship of natural resources, particularly challenging the notion of a farming metaphor for marine capture fisheries. The next chapter then provides a simple three-step framework from which we can begin to provide and utilize advice on living marine resource (LMR) management.

So that none of this is viewed as a collection of isolated theories, even more rambling about management frameworks, or as related arcane musings, the final chapter in this section then briefly elucidates specific case studies where an EBFM has been warranted or even begun to be implemented. Those examples are presented to demonstrate a clear need for EBFM.

1

Admit the problem

Houston, we have a problem.

Jim Lovell, Apollo 13 mission commander (attributed)

THE SCOPE AND EXTENT OF GLOBAL FISHERIES ISSUES

There is a global crisis in marine capture fisheries that has been punctuated by a few success stories. This crisis has been expressed in several ways, chief of which is the collapse of targeted stocks (sensu Botsford *et al.* 1997; NRC 1999; Jackson *et al.* 2001; Pitcher 2001; Pauly *et al.* 2002, 2003; Garcia and de Leiva Moreno 2003; Rosenberg 2003; Beddington and Kirkwood 2005; Mullon *et al.* 2005; Berkes *et al.* 2006; Beddington *et al.* 2007; FAO 2007). Both the global catch and the vast majority of the world's fish standing stock biomasses either have plateaued or are down. Specifically, over ~70% of the world's fish stocks are at overfished or fully utilized levels (NRC 1999; FAO 2007; Figure 1.1). We can argue over the finer points regarding whether we have 90% or 40% of the "virgin" biomass of many of these species (e.g. Myers and Worm 2003; Sibert *et al.* 2006; Hilborn 2006), but the germane point is that we have removed a significant fraction of these fish stocks. It is unclear whether these stocks can recover and if the associated impacts from their overfishing can be overcome (sensu Hutchings 2000; Pauly *et al.* 2002; Hutchings and Reynolds 2004).

Most of the larger-sized fish in the world's oceans are much less abundant than they were 50 years ago (Myers and Worm 2003; Worm *et al.* 2005; Sibert *et al.* 2006; Figure 1.2). This refers both to those species that are generally larger and to individuals within a species, the latter a classic and common sign of growth overfishing, an event that has been exhibited regularly. There are today generally smaller-sized species

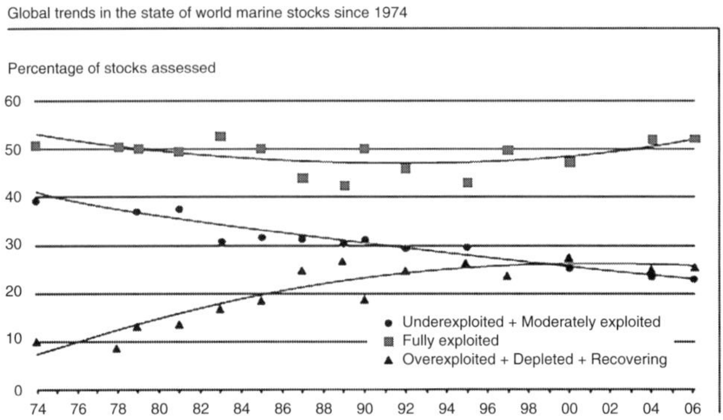

Figure 1.1 Global status of fisheries stocks. (Adapted from FAO 2007.)

dominating the fish communities of the world's oceans, with fewer highly migratory species (e.g. tunas, billfishes) than there were 50 years ago. There are mainly smaller-sized individuals among those species, evidenced by the increasingly rare anecdotal reports of "whale cod" or "monster halibut" on the world's fishing docks.

There is growing evidence that we are "fishing down the food web", or at least through it, by removing upper trophic level predators or the functional redundancies in a system (Pauly *et al.* 1998, 2002; Essington *et al.* 2006; Figure 1.3). This includes not only large fishes like tunas, swordfishes, or sharks, but also many marine mammals (Jackson *et al.* 2001; Lotze *et al.* 2006). Conversely, there is a notable increase in small pelagic fishes and their associated fisheries, as well as a noted increase in invertebrate fisheries. Will we begin prosecuting fisheries for gelatinous zooplankton in the near future? This seemingly ridiculous comment is not as far from reality as it might originally seem (Pauly *et al.* 1998, 2002, 2003; D. Pauly, personal communication, 2003). In many oceans, seas, and large bodies of water, the population explosion of both gelatinous organisms and other invertebrates is not uncommon (e.g. Purcell *et al.* 2001; Shiganova *et al.* 2001; Brodeur *et al.* 2002; Gucu 2002; Worm and Myers 2003; Bilio and Niermann 2004).

Further, there have been numerous instances suggestive of recruitment failures in the world's fisheries (e.g. Myers *et al.* 1997; Beverton 1998; Hutchings 2000; Walters and Kitchell 2001). It has been hypothesized that many fisheries have been perturbed beyond their capacity to absorb any further impact to spawning biomass and recruitment potential. More disturbing, few stocks actually exhibit patterns that match the standard recruitment models of Ricker or Beverton–Holt, models which at least partially form the basis for

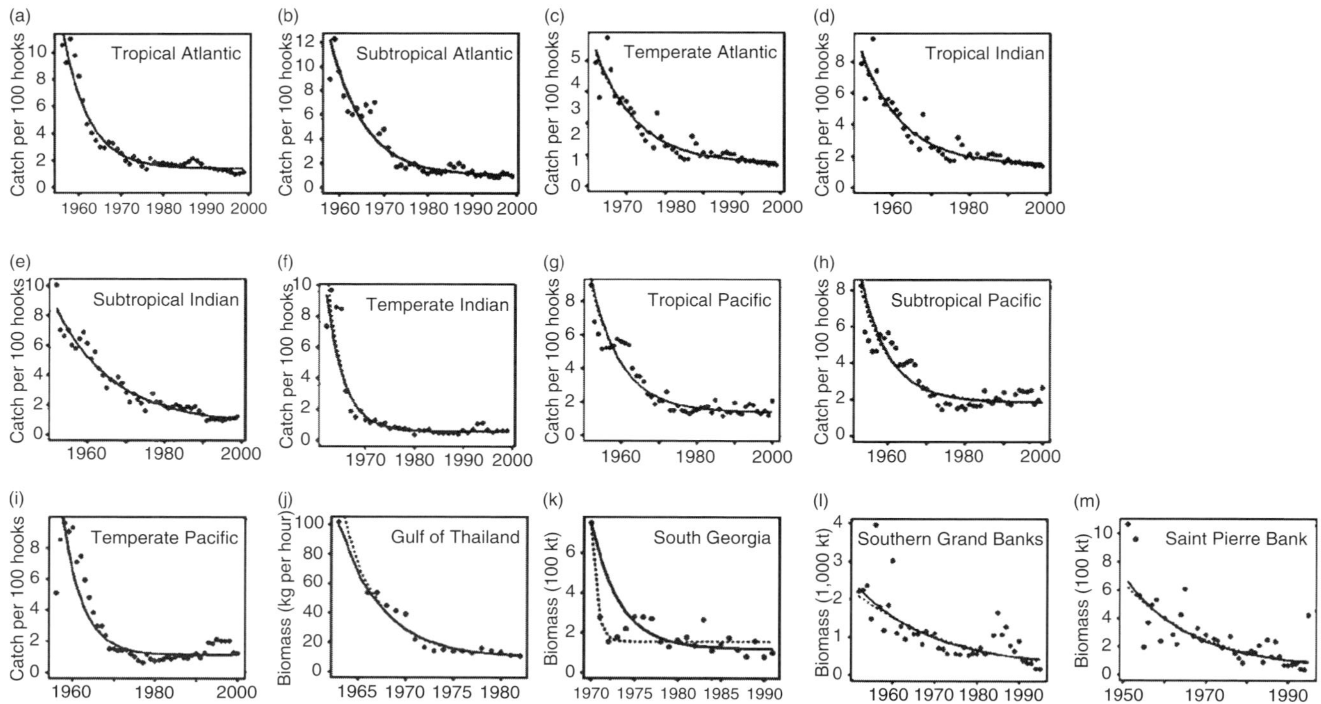

Figure 1.2 Trends in biomass or catch of big, predatory fish. (Adapted from Myers and Worm 2003; see also Worm *et al.* 2005.) Despite caveats over the use of catch data (versus fishery independent sampling), despite omission of some examples showing different (flat or positive) trends, and despite caveats over 40 or 90 or whatever percentage depletion, it is understood that there are fewer larger fish in the world's oceans.

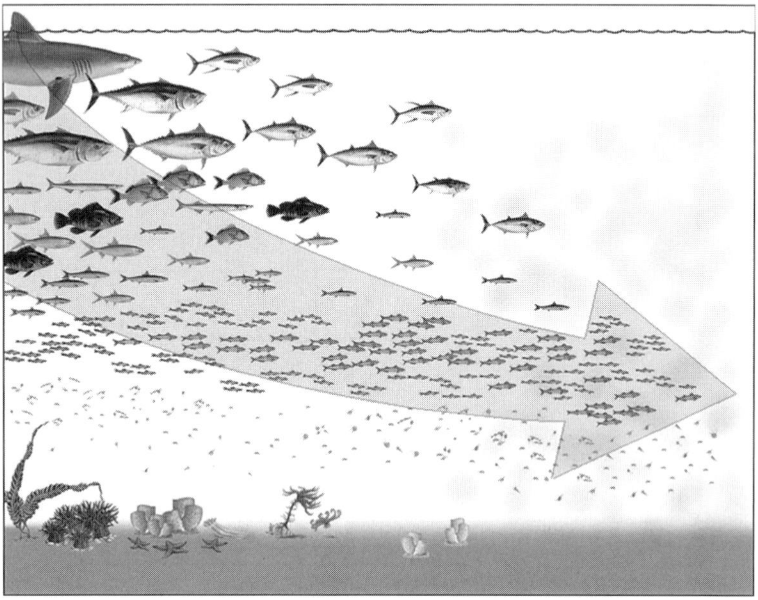

Figure 1.3 Diagram of fishing down the food web. (Adapted from Pauly 2003.) Despite some caveats of fishing down versus fish through a food web (Essington *et al.* 2006) and despite some regions specifically and intentionally targeting benthos and small pelagics, it is understood conceptually that fisheries are generally now fishing on organisms that are at lower trophic levels than levels at which they used to fish.

management of many of the world's fisheries. More telling, after recruitment overfishing has occurred, few stocks have exhibited recoveries to the level of stock biomass that demonstrate the near levels of unexploited recruitment and biomass possible for these assumed curves (sensu Hutchings 2000; Hutchings and Reynolds 2004; Figure 1.4). Fishing might even be impacting the genetics and, subsequently, the evolutionary responses of fish populations (Conover and Munch 2002).

We also see multiple evidences of sequential depletion, whereby a fleet depletes a stock, the management responds too late to mitigate the depletion, the fleet moves on to target another stock, and so on, ad infinitum. Many of the world's premier invertebrate fisheries developed in response to the decline in groundfish stocks (e.g. Pauly *et al.* 1998; Worm and Myers 2003). Some of the leading fisheries today target stocks that 20 to 30 years ago were considered trash or bycatch species (Link 2007).

The above considerations do not include all of the potentially negative effects of fishing on the environment. Concerns about mobile

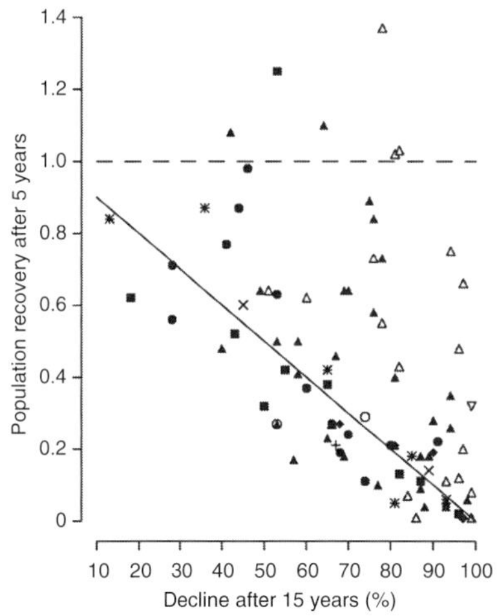

Figure 1.4 Lack of recovery in multiple fish populations, largely attributed to failed recruitment. (Adapted from Hutchings 2000.) Slanted crosses, Engraulidae; upward triangles, Clupeidae; downward triangles, Osmeridae; filled triangles, Gadidae; stars, Scorpaenidae; upright crosses, Anoplopomatidae; filled diamonds, Sparidae; diamonds, Nototheniidae; filled squares, Scombridae; filled circles, Pleuronectidae; circles, Soleidae.

bottom-tending fishery gear, as it affects fish habitats in particular and benthic habitats in general, are widespread (Figure 1.5; Sainsbury 1988; Collie *et al.* 1997, 2000; Jennings and Kaiser 1998). Bycatch and impacts on protected, endangered, and threatened species (PETS) and non-target species (NTS) are also pervasive. Another major question is how the fundamental properties of ecosystems (biomass structure, energy flows, hysteresis, etc.) are altered in response to fisheries. Similarly unknown is how the interconnections and interdependencies among species are altered (e.g. Jennings and Kaiser 1998; Scheffer *et al.* 2005; Myers *et al.* 2007).

The point of raising these issues is to highlight that there is a major, pandemic problem in marine capture fisheries.

ADMITTING WE HAVE A PROBLEM WITH GLOBAL FISHERIES

The global marine fisheries predicament is not trivial, has huge socioeconomic implications, and is unlikely to be addressed under the status quo approaches (e.g. Larkin 1996; Pitcher 2001; Link 2002b, 2002c). There are three false assumptions about the global fisheries problem which have persisted, however, and they merit examining before we proceed.

(a) (b)

Figure 1.5 Pictures from the bottom of Georges Bank in areas that have been closed (a) and open (b) to fishing. (Adapted from Collie *et al.* 1997.) The difference in biogenic habitat is not subtle.

The first false assumption is that there are no success stories at all. By contrast, there have been some clear and celebrated instances of stock recovery. For instance, three examples that come immediately to mind are the striped bass along the US East Coast (Richards and Rago 1999), the lake trout in the Laurentian Great Lakes (Selgeby *et al.* 1995), and Georges Bank scallops (Murawski *et al.* 2000). The walleye pollock fishery in the Gulf of Alaska is a well-managed fishery (Hollowed *et al.* 1997, 2000b; B. Megrey,

Graham's Law of Overfishing

C ↓

E ↑

repeat

Figure 1.6 Schematic of Graham's Law of Overfishing (sensu Graham 1943). C, catch; E, effort.

personal communication, 2005), as are various prawn fisheries in Australia (Dichmont *et al.* 2006a, 2006b; A. Smith, personal communication, 2004). There are more examples, but the point is that the proverbial sky is not falling, at least not yet, but it is certainly changing color. Not every fishery is entirely bungled up, and we need to be honest about this fact.

Second is the assumption that the oceans and their inherent productive capacity (i.e. their ability to produce biomass of interest for human consumption) have changed. In many instances we have seen tradeoffs in fundamental types of species prosecuted for fisheries, but we have not necessarily irrevocably hampered the productive capacity of the world's oceans (Pauly and Christensen 1995; Ware and Thomson 2005) – at least not yet (Brander 2007). That is, on the basis of sunlight and nutrients – with all the appropriate caveats of physical dynamics, upwellings, large-scale climate drivers (e.g. Pacific Decadal Oscillation [PDO], El Nino Southern Oscillation [ENSO], Atlantic Multidecadal Oscillation [AMO], North Atlantic Oscillation [NAO], etc.) – it is unlikely that the amounts of biological production capable in any particular system have fundamentally changed beyond the range of natural variation. Rather, how that production is allocated has changed greatly. In many respects, this point is the major theme of this book.

Finally, the third false assumption is that everyone agrees or understands that there is a major problem. In many instances this is not the case. Both the world's fishing industries and fisheries management agencies have institutional inertia – this needs to be overcome for this problem to be better addressed and to avoid further negative outcomes as predicted from Graham's Law (Figure 1.6; Graham 1943; Smith 1994). Often there is unawareness or denial that the predicament is as pervasive as it is. We will remain unable to address the problem until we understand: (1) we have a problem; (2) how we got into the current crisis in the first place; and (3) we ought to explore some steps to mitigate the predicament we are in.

Denying the severity of the issue is not helpful, but neither is overstating or misstating its seriousness. Certainly the problem sells lots of headlines and is a hot topic even in major scientific journals, but discussing the matter here in a balanced, objective, and pragmatic manner is my goal. Very simply, the aim of this book is to explore possible ways that we can address this issue, particularly by examining fisheries management in a more holistic, broader context.

HOW DID WE GET HERE?

Those who cannot remember the past are condemned to repeat it.

George Santayana, *Reason in Common Sense*

Much of the current global fisheries predicament is not surprising given Graham's Law of Overfishing (Graham 1943; Smith 1994; Figure 1.6). Graham's Law effectively states that short of any controls or intervention, the typical pattern for a fishery is to increase catch, which lowers standing stock biomass, which leads to an increase in effort, which leads to an increase in catch, and so on, the cycle iterating until a stock collapses. This pattern has been repeatedly noted for fish stocks the world over (Hilborn and Walters 1992; Beverton 1998; Pauly *et al.* 2002, 2003; Rosenberg 2003; Beddington and Kirkwood 2005; Link 2007). A corollary to Graham's Law might be called the Law of Sequential Depletion. We see this in many instances, whereby the primary targeted stock collapses (i.e. can no longer support a commercial fishery) and the fleet then moves on to target another stock, and so on, ad infinitum.

For example, a hundred years ago in US waters of the North Atlantic, the situation looked like this: the Atlantic halibut fishery was blissfully operating without any idea that in less than two decades it would collapse; the cod fishery was maintaining its ability to catch fish; the redfish and haddock fisheries were just being developed or were ramping up; spiny dogfish and goosefish were nuisance bycatch species at best; and lobster was considered such trash that it was used to fertilize gardens. In fact, there are stories of children being hounded mercilessly at school for bringing a lobster for lunch. There are even records of prison riots because lobster was served three times in one week; imagine the ignobility of it all (sensu Smith 1994; T. D. Smith, personal communication, 1999)!

The fisheries scene looks very different in the present day. Goosefish is now the most valuable finfish fishery in the region (at about US $30–$35 million), coming after a noted decline in the major spiny dogfish fishery (DOC 2005). The major gadid and flatfish species (e.g. cod, haddock, and several flounders) are still important culturally but amount to a fraction of the value of landings in this region. Even fisheries of goosefish, the most valuable finfish, are minor compared with most fisheries of invertebrates in the region. Two species of squid and surf clams, which are both minor fisheries in terms of total biomass removed, are valued at the same dollar amount as the most valuable finfish fishery. Sea scallops are ten times more valuable, and the lobster, which only one century ago was so highly despised, now accounts for greater than $300 million in value of landings (DOC 2005). This example is illustrative of the global phenomenon that as fish stocks are overexploited, fishing fleets pursue new species.

So how did we end up here? Let me provide an overview of fishing, fisheries management, and fisheries science for the past 100 years or so.

In the time period from 1850 to 1899, the emphasis of marine, aquatic, and fish-related science was on natural history. By its very nature, this was an inclusive, holistic, observational approach. It was also distinctly nonquantitative. Simultaneously, many of the first national marine or fisheries laboratories were formed, highlighting for the first time government-subsidized biological research on a large scale. In many ways these labs were established to provide information in the national or regional interest and to mediate differences among states and nations. The technical element of these endeavors was moving beyond an accumulation of species lists towards an understanding that serious declines in living marine resources were being observed, and prominent members of society (i.e. those who had a financial stake in those resources) wanted to know why. It was at this point that a decline in many marine organism populations was truly recognized (sensu Smith 1994). By the beginning of the 1900s, there was a paradigm shift that living marine resources might not be inexhaustible, in stark contrast to the common paraphrase of Thomas Huxley's (1884) famous quote, "The world's fish and fisheries are inexhaustible."

Over the next 20 years many biological fields of study emerged as formal scientific disciplines. This began a half century of "physics envy" that led to increasingly quantitative and reductionist approaches to marine and fisheries science. Concurrently, a notable phenomenon was observed regarding fish stocks: it was obvious that there was a major change in North Sea stocks during and after World War I (e.g. Lotka 1925; Volterra 1926). Attempting to determine cause and effect relationships for this observation led to a wide range of political, economic, and similar pressures associated with fisheries management and fisheries science, to the point that those issues escalated into major disputes. In fact, several international disputes and scientific debates intensified during that era and in many respects continue to this day. For instance, arguments over whether fishing or environmental factors (or both) were the primary causes of stock declines were brought to the fore (e.g. the Thompson–Burkenroad debate; see Smith 1994 for a full description). Despite this backdrop, there were still relatively simple fishing methods being employed at this time due the cost ineffectiveness of more powerful engines for most fishing fleets.

From 1920 to 1940, we saw the formulation of key biological processes. Lotka–Volterra models, Elton's pyramid, Gaussian models, Hjort's model, Pearl's model, Clements and Shelford discussions, etc., all of which sought to provide a mathematical representation of basic biology – essentially growth and death with various specialized denotations thereof. Many of these were developed independently, arrived at similar formulations of processes, and effectively confirmed the commonality and generality of these mathematical representations. Yet due to the more reductionist approach generally employed, coupled with other social considerations (the postindustrialization boom which still provided a strong impetus for mechanistic thinking; recovery from the Great Depression and the individualized self-sufficiency mentality it fostered; less rapid forms of communication compared with today, etc.), there arose a disciplinary specialization among those endeavors of investigation commonly associated with fisheries issues. In particular, ecology, oceanography, and fisheries biology became distinct fields.

During this period fishing technology rapidly changed. Two particular innovations irrevocably altered the seascape of fishing that had mostly remained remarkably unchanged for millennia. The advent of on-site freezing was a major invention that obviously altered how fleets processed their catches (Smith 1994). Another big change was that the majority of the world's fishing fleets converted to an industrialized operation, reflective of what had occurred in agriculture about 50 years earlier. Generally speaking, fleets converted from sail to engine-powered vessels and from longlining/handlining to bottom trawls. These developments, coupled with on-site freezing, allowed fishing vessels to stay at sea longer, develop larger holds, catch more fish per trip, and ultimately increase landings. This then led to fish no longer being thought of as an inexhaustible resource. Further, as these new technologies developed, stocks that were previously unavailable or generally not targeted became more viable fisheries to prosecute, thus opening up new subsectors within the overall fishing industry.

The next 20 years (from 1940 to 1960) was the main time frame from which the forebearers of current fisheries management and scientific approaches can be directly traced. There was a clear single species emphasis in fisheries science, with models and approaches developed from Graham, Ricker, Beverton and Holt, Schaefer, Leslie, and the like, and with slight modifications they are used in many of today's stock assessments. This approach reflected the reductionist and mechanistic thinking that was prevalent at the time. Conversely, there was a community (multispecies [MS] in fisheries parlance) emphasis in ecology during this time. The debates over density dependent versus independent factors were raging, with seminal works by Andrewartha and Birch, Lindeman, Holling, and Hutchinson focusing on how species interacted with their environment and one another. Also at this time were rapid developments in oceanography, with the first attempts at quantification of key physical and chemical processes on large scales. Interestingly, in many respects marine biology still had a naturalist focus (e.g. rocky intertidal). Although my characterization is oversimplified, one can see how the disciplinary specialization led to thinking that produced distinct and disparate views of how the oceans and their living marine resources functioned. Meanwhile, as in the previous world war, notable changes in fish stocks were observed during and after World War II. By this point, the relationship between cessation of fishing and recovery of fish stocks was less debatable than it had been 40 years earlier (Smith 1994). Once fishing activities resumed, however, disputes over who was causing the most problems and who should be allowed to catch what fraction of a stock rapidly escalated. Many international living marine resource (LMR) management advice-producing organizations (e.g. International Commission for the Northwest Atlantic Fisheries [ICNAF], International Council for the Exploration of the Sea [ICES], International Whaling Commission [IWC], International Commission for the Conservation of Atlantic Tunas [ICCAT], Food and Agriculture Organization [FAO], etc.) were formed or bolstered during this era, with much of their advice and institutional processes rigorously formalized.

In the 1960s to early 1980s, a multispecies emphasis in fisheries models was developed that recognized the importance of both species interactions and the ubiquity of bycatch. In one form or another (cf. Chapter 7; Daan and Sissenwine 1991; Hollowed *et al.* 2001; Whipple *et al.* 2000), multiple species were being considered concurrently. In most cases, however, this was a research emphasis and most LMR management advice was still focused on single species (which continues to this day). Many of the LMR management institutions had built up processes and machineries such that they could not adapt to this additional information. In particular, often the multispecies information was more conservative than the single species advice, and as such (i.e. less fish able to be caught) was typically rejected out of hand for political reasons. During this time, the perception was that the academic or even somewhat esoteric discipline of ecology left the ivory tower to focus on major, global, and highly visible environmental problems, albeit mainly for terrestrial or inland water issues. Concerns about contaminants, organic pollution, world population growth, and so forth led to programs like the International Biological Program (IBP) and the development of systems ecology. Meanwhile, there was an emerging physical–biological emphasis in oceanography that started to minimize the distinction between various forms (organic or inorganic) of carbon and nitrogen and recognized that various processes were all usually part of one bigger cycle. During this era, conservation awareness was reinvigorated such that, at least in the USA, passage of environmental-protecting legislation was unparalleled in all of human history. The Magnuson Act, ESA, MMPA, NEPA, CERCLA, AQA, WQA, UNCOLS (Law of the Sea), and dozens of others like them were instituted. Germane to fisheries was the establishment of economic exclusive zones (EEZ), which effected the nationalization of fisheries for most fishing grounds, save the high seas.

From the 1980s to present, we have entered an era of synthesis and integration. We see the fusion of ecological theories (food web theory, trophic cascades, landscape ecology, etc.); the attempts at global integration for many oceanographic factors (particularly with the advent of remote sensing, automated data collection, satellite imagery, large-scale telemetry, etc.); and we see the extension of fisheries models to incorporate more multispecies models, habitat concerns, ecosystem models, etc. (Daan and Sissenwine 1991; Hollowed *et al.* 2000a; Whipple *et al.* 2000; Plagányi 2007). Thus in many respects the contemporary era represents a reconvergence of different disciplines. Simultaneously, society is recognizing the truly global nature of human impacts on the environment, with particular concerns over climate change and induced global warming (IPCC 2007). A recognition that we need to consider the interconnectedness of issues and examine more integrated, holistic views of how systems operate is returning. All the while, overfishing continues. Again, over 70% of world stocks are at or near overfished levels (FAO 2007) and sequential depletion of fish stocks occurs regularly (Pauly *et al.* 1998; Link 2007).

Given this general history, it is easy to note some key developments in fishing practices, the fishing industry, and fisheries science. Yet in many ways fisheries management remains fundamentally the same today as it did 50 years ago, if not even 150 years ago. Smart individuals from prior centuries noted these problems and our need to sift through them. For instance, this quote by Spencer Baird (1873), the founder of the US Fisheries Service, describes the possible causes for declines in New England fish stocks:

> (1) The decrease or disappearance of the food upon which the fish subsist, necessitating their departure to other localities.
> (2) A change of location, either entirely capricious or induced by the necessity of looking for food elsewhere, as just referred to.
> (3) Epidemic diseases, or peculiar atmospheric agencies, such as heat, cold, etc.
> (4) Destruction by other fishes.
> (5) The agency of man; this being manifested either in the pollution of the water by the discharge into it of the refuse of manufactories, etc. or by excessive overfishing, or the use of improper apparatus.

This nineteenth-century quote is as illustrative as it is haunting in that although the particular buzzwords have changed, many of the fundamental concepts and issues facing us have not been resolved. The main point is that Graham's Law of Overfishing has predicted the crises we face, but Baird and others long ago noted that we need to explore the issue of fisheries in a broader context.

AN EMPHASIS ON SINGLE SPECIES

The science behind fisheries has been decidedly single species (SS), or more precisely, with a stock or population emphasis. In particular, the emphasis in much of applied fisheries science is on the refinement of stock assessments and associated methodologies. The four main categories of assessment approaches are: index-based, recruitment-based, production-based, or cohort-based. The main types of management advice from these approaches have generally been of the type that keys off abundance (N), recruitment (R), biomass (B, particularly spawning stock biomass, SSB) and/or fishing mortality (F). A lot of time and effort go into these estimates, with an ever-increasing demand for enhanced precision.

Similarly, the management of fisheries and more broadly all LMRs (i.e. including marine mammals) has been for the most part single species in orientation. Institutional structures the world over generally seek to manage fish stocks on a stock-by-stock basis; this is a logical and natural product of the underlying scientific models used for most fisheries assessments. Using the assessment advice to nominally determine the status of stocks relative to some established benchmark levels (i.e. reference points in fisheries parlance), decision criteria, or control rules are then implemented accordingly. In practice, usually estimates of *B* or *F* are compared with some limits or thresholds and then fishing effort is adjusted accordingly (usually down) via several fishery management tools or measures (which I will describe in later chapters). In theory, once the levels of *B* and/or *F* are set and the adjustments that need to be made are noted, allocations across a fishing sector are then negotiated. In practice, this has been difficult (for obvious reasons), such that *B* or *F* are often revisited in the course of attempting to allocate what the actual allowable catches or effort are for a fleet.

Many have espoused that if we just did single-species management correctly we would be in good shape (e.g. Beverton 1998; Hilborn 2004; Mace 2004; see perspective of Eagle in Leslie *et al.* 2008). There is a clear recognition that the failure by many LMR management institutions to heed the scientific advice (e.g. Rosenberg *et al.* 1993; Beverton 1998; Caddy 1999; Caddy and Cochrane 2001; Sissenwine and Mace 2003; Caddy and Agnew 2004) has led to many of the fisheries problems we see today. These are all fair points. In fact, in his highly useful book on the community and ecosystem effects of fishing, it is telling that Hall (1999) has one ending chapter titled "Reduce effort, reduce effort, reduce effort," an interesting conclusion given the breadth and scope of all the topics in that book. However, even if we did LMR management 100% according to the SS scientific advice, I and several others would argue that there are still issues remaining which we would not be able to address (e.g. Larkin 1996; Botsford *et al.* 1997; Link 2002b, 2002c; Pikitch *et al.* 2004; see perspectives of Cury (2004); Lotze (2004); Field and Francis 2006; Murawski 2007).

Interestingly, the fisher communities and fisheries built up around various fish stocks are at least multispecies in terms of what is caught and often what is landed. It is market structures (in the economic sense) that drive which stocks are prosecuted and when, but the actual catch is decidedly multispecies, which the market has usually been able to absorb. In fact, a multispecies fisheries "portfolio" has several desirable and robust economic properties (Edwards *et al.* 2004). However, the majority of the world's overriding policies influencing these fisheries are based on single-species management and its supporting science.

By default, all these underlying assumptions assume that processes internal to a stock and fishing mortality are by far the largest drivers of fish population dynamics. In some instances, this may in fact not be the case.

TRANSITIONS

We have noted that single-species (SS) approaches to LMRs can provide very useful insights into population dynamics of a stock. Factors such as age- or size-based variation in demographic parameters, growth rates, stock–recruitment relationships, density-dependent effects, and mortality are readily estimable in most LMR management and scientific contexts. Insights from elucidating and understanding these dynamics are highly useful, and in many ways, determining the status of stocks is the cornerstone of fisheries science. Knowing the status of these stocks relative to key benchmarks will, in many instances by law, continue to be a necessity for LMR management in many countries.

However, SS approaches do have their limitations. As they are currently used, they will rarely provide insights nor the ability to address how a particular stock might be impacted by other factors such as species interactions, predation mortality, forage limitations to growth or recruitment, thermal limitations to distribution, and other environmental constraints to growth (or distribution, or recruitment, or other production considerations, etc.). Beyond the stock focus, SS approaches will never be able to provide insights into a broader suite of pressing issues: for example, changes in ecosystem structure and functioning, biodiversity, fishing gear impacts on habitat, needs of protected or rare species relative to other species or other sources of removals (e.g. fishing), ecosystem effects of discarding unwanted bycatch, fishing impacts on energy flows of a food web, or particularly non-fishing but LMR-related ecosystem services and values, and so on. As has been the case off and on for the past 150 years, debates are resurgent over which of these processes are most important in determining the fate of LMR populations and communities.

So, why do we need a broader, more holistic (ecosystem-based) approach beyond typical SS approaches that are mostly used in contemporary LMR management? There are four main reasons.

First, these broader approaches address (or at least should attempt to assess) the effects of fishing on non-target species, protected species, habitat, ecological interactions, and system-wide processes. This is not a trivial endeavor, but in the USA in the National Environmental

Policy Act (NEPA) rubric (and similar environmental laws in other countries) it is required to allow for or mitigate any activity that could substantially alter or impact the environment, which includes fishing activities. Formerly, these NEPA or environmental-impact types of considerations had not been broadly applied (or were only done in minimal terms) for large marine ecosystems such as continental shelves, as these locales were understood to have been supporting fishing activities for centuries. However, recent legal activities and other ocean use cross-sector conflicts have brought this issue to the fore, such that now a more comprehensive, holistic consideration of all the potential effects of fishing is becoming more and more routine. It has not reached the point of reversing the burden of proof regarding allowing fishing or any other activity just yet, but several countries and certification programs are moving towards that direction (sensu Dayton 1998; Kaiser and Edwards-Jones 2006).

Second, these broader approaches recognize that marine ecosystems provide "goods and services" other than fishery harvest; several sectors have developed to exploit or utilize these resources. A list can be developed from first principles and all are sectors with clear and measurable economic benefits to society: mineral extraction (gas, oil, etc.), energy production (wind farms, tidal turbines, etc.), navigation and transportation, tourism (SCUBA diving, recreating, sunbathing, etc.), ecotourism (e.g. whale watching), novel medicine development, coastal zone development, etc. The world's oceans also provide services that are difficult to value economically. Considerations such as carbon dioxide scrubbing and fixation, other atmospheric gaseous exchanges, climate and weather forcing, etc., and the simple aesthetic appreciation most people have of the ocean are clearly key functions of the ocean, even if nearly impossible to tag with a dollar value. Without getting into full-blown ecosystem approaches to management across multiple use sectors (see Chapter 2), the point is that these other sectors can impact and be impacted by the fisheries, and more generally, the LMR sector. Accounting for these potential cross-sector interactions is critical.

Third, these ecosystem-based approaches arguably provide more accurate assessments and evaluations of the dynamics of LMR populations and communities. I have seen a SS assessment go through the process of painstakingly validating the rigor of its precision and model outputs, such that the fishing removals were at a level of *X*, the stock was at a value of *Y* and the projections were for *Z*. The model performance was not "converging" (i.e. behaving as expected, fitting the input data, or displaying acceptable diagnostics), and the overall assessment was generally unsatisfactory. This proved to be a difficult and frustrating situation for all involved, as it seemed that there was a bit of information missing. Of course, in the SS framework it was difficult to include another source of removals (in this particular case due to predation mortality) which would have accounted for an additional 50 to 500*X* of removals, notably altered estimates of *Y* and *Z*, and improved the overall assessment of this particular stock. This example (one among many) illustrates the point that incorporating broader ecosystem considerations can improve scientific information, enhance management advice, and provide the basis for more sustainable decision criteria.

Finally, these broader approaches explicitly address biomass tradeoffs. What biological configuration would we like for a marine ecosystem, within the limits of biological, ecological, and energetic feasibility? There is only a limited amount of primary production for any given ecosystem; how will it be apportioned into LMR biomass? To be sober, the sum of single species biomass at maximum sustainable yield (B_{MSY}) is greater than B_{MSY} for a particular system, and it may be energetically impossible to simultaneously maximize yield for multiple species (Brown *et al.* 1976; May *et al.* 1979). Larkin's (1996) harbinger discussion on the topic narrowed in on this as *the* key objective of LMR management.

As difficult as it may be, we should seek to specify the species mix desired in the fish assemblage of an ecosystem. This final point is the crux of the matter. Thus the choice of subtitle for this book is "Confronting tradeoffs". Without confronting tradeoffs directly, we will never be able to directly do EBFM. Doing EBFM is all about confronting tradeoffs.

SUMMARY POINTS

- The world's fisheries and living marine resources (LMRs) in general are facing some significant challenges.
- Single species (SS) approaches are and will remain useful, but they will not solve all of the problems facing LMR management.
- Ecosystem-based fisheries management (EBFM) recognizes the effects of fishing on other aspects of marine ecosystems beyond direct impacts to LMRs. It also recognizes that there are other uses of the ocean besides fishing.
- Ecosystem considerations can provide improved information and assessments for the management of LMRs.
- EBFM explicitly addresses biomass tradeoffs.
- There is a need to implement EBFM.

2

Why is an ecosystem approach now strongly heralded and merited?

> Ecosystems are not more complicated than you think, they are more complicated than you *can* think.
>
> Jack Ward Thomas, former director of the US Forest Service (Burnside and Rasmussen 1997)

SORTING OUT ALPHABET SOUP, AND WHAT DOES EBFM MEAN?

Ecosystem considerations in a marine scientific and management context have been extant for more than a century (e.g. Baird 1873). The concept of ecosystem-based fisheries management (EBFM) as applied to marine fisheries was more recently crystallized in Larkin's (1996) paper, with the chief observation that it provides a holistic approach to natural resource management. Since then, there has been recognition that EBFM for marine systems is gaining momentum (Botsford *et al.* 1997; NMFS 1999; Link 2002b, 2002c; FAO 2003; Garcia *et al.* 2003; Pikitch *et al.* 2004; see perspectives of Cury 2004; Lotze 2004; Browman and Stergiou 2005; Garcia and Cochrane 2005; Field and Francis 2006; Marasco *et al.* 2007; Murawski 2007). Most who have thought about the issue conclude that we can only manage *human activities* within an ecosystem (Hilborn 2007), not necessarily the ecosystem itself. The key point is that this and related terminology are, as Larkin noted, shorthand for more holistic approaches to resource management. There has been no shortage of attempts to define ecosystem management, for both marine (e.g. Larkin 1996; Botsford *et al.* 1997; NMFS 1999; Link 2002b; FAO 2003; Garcia *et al.* 2003; Browman and Stergiou 2004; Pikitch *et al.* 2004; POC 2004; USCOP 2004; Garcia and Cochrane 2005) and terrestrial ecosystems (or generally speaking any ecosystem; Slocombe 1993; Grumbine 1994, 1997; Christensen *et al.* 1996; Haeuber 1996; Yaffee 1996, 1999; Yaffee *et al.* 1996; Franklin 1997; Lackey 1998). Some coordination or organization thereof would seem to be necessary.

Given all the terms in use on this topic, it would be helpful to clarify them. I am also unaware of any school that deciphers all the extant institutional acronyms in use on this topic. Thus, allow me to list some common ecosystem-related terms or phrases before discussing the concepts behind them (Table 2.1).

First, there is *ecosystem management* (EM). Technically this is impossible, as we are in reality managing human activities in any ecosystem and not the weather, species interactions, etc. Given this caveat, ecosystem management is understood to be a moniker for more broadly considered management. It has come to mean a catch-all term for the general management of natural resources in a more holistic, sometimes amorphous manner. The problem with this term is that it can lead to other terms such as "ecosystem integrity" or "ecosystem health" and similar such ambiguities (Wicklum and Davies 1995; Link 2002c), none of which lend themselves to measurability, repeatability, or evaluation of specified processes over time. As shorthand for more holistic resource management approaches, EM is sensible. As an operational or technical definition, it poses problems.

Following on are the more technically accurate terms *ecosystem-based management* (EBM) and *ecosystem approaches to management* (EAM). EBM implies the application of management options to a resource sector that is holistic and integrated, accounting for all key factors and processes from the primary perspective of the entire (eco)system. I tend to think of this as starting with the entire ecosystem and then working down into each of the sectors as they interact among one another. Conversely, EAM is doing within-sector resource management cognizant of the effects of other sectors on each other. I consider EAM to be starting within a sector and working up to include other potential factors arising from other sectors. In practice, EAM is rarely distinguished from EBM. More germane, in marine resource circles, these two terms typically invoke concepts like multiple use management (MUM), regional ecosystem management (REM), ocean zoning (OZ), and integrated management (IM), with the explicit understanding that all involved sectors are being considered and consulted.

Taking EBM and EAM to the specific sector of fisheries produces two more terms: *ecosystem approaches to fisheries* (EAF or EAFM) and *ecosystem-based fisheries management* (EBFM). The first aims to do (within-sector) SS fisheries management cognizant of the effects of other sectors on fisheries, the effects of fisheries on other sectors, and perhaps the effects of fisheries on the ecosystem (e.g. gear impacts on benthic habitats). Conversely, EBFM aims to manage fisheries by coordinating,

Table 2.1. *The scope and implications of various terms for ecosystem approaches to living marine resource management*

Often these terms are used interchangeably, and despite subtle distinctions, most individuals recognize that they are effectively shorthand for implying more holistic approaches to resource management.

Title	Term	Connotations
EM	Ecosystem management	A catch-all term for the general management of natural resources in a more holistic, sometimes amorphous manner. It is recognized that technically this is infeasible as we are really managing human activities in any ecosystem, but it is understood to be a moniker for more broadly considered management.
EBM	Ecosystem-based management (a.k.a. MUM, REM, OZ, IM, etc.)	Applying management options to a resource sector that is holistic and integrated, accounting for all key factors and processes from the primary perspective of the entire (eco)system. In marine resource circles, this is typically invoking concepts like multiple use management or regional ecosystem management of ocean zoning or integrated management, with the explicit understanding that all involved sectors are being considered and consulted. Think of this as starting with the entire ecosystem and then working down into each of the sectors as they interact among one another.
EAM	Ecosystem approaches to management	Doing within-sector resource management while cognizant of the effects of other sectors on each other. In practice, EAM is rarely distinguished from EBM. Think of this as starting within a sector and working up to include other potential factors arising from other sectors.

EAF (EAFM)	Ecosystem approaches to fisheries (management)	Doing within-sector single-species fisheries management cognizant of the effects of other sectors on fisheries and the effects of fisheries on other sectors and perhaps the effects of fisheries on the ecosystem (e.g. gear impacts on benthic habitats). Although mildly distinct from EBFM, the two are used interchangeably. In EAF, though fisheries are emphasized, the inclusion of marine mammals, protected species, non-target species, etc. is usually implied. Think of starting with a stock and building up to include as many factors as are appropriate. In Figure 2.1, build from the left and work up the gradient.
EBFM	Ecosystem-based fisheries management	Managing fisheries to coordinate, account for, and include all factors in a holistic, synthetic, integrated fashion. Fisheries are emphasized, but the inclusion of marine mammals, protected species, non-target species, etc., are explicitly included. Although mildly distinct from EAF, the two are used interchangeably. Think of this as starting with an ecosystem perspective of fisheries production that addresses specific interactions among species and with the environment. In Figure 2.1, build from the right and work down the gradient.

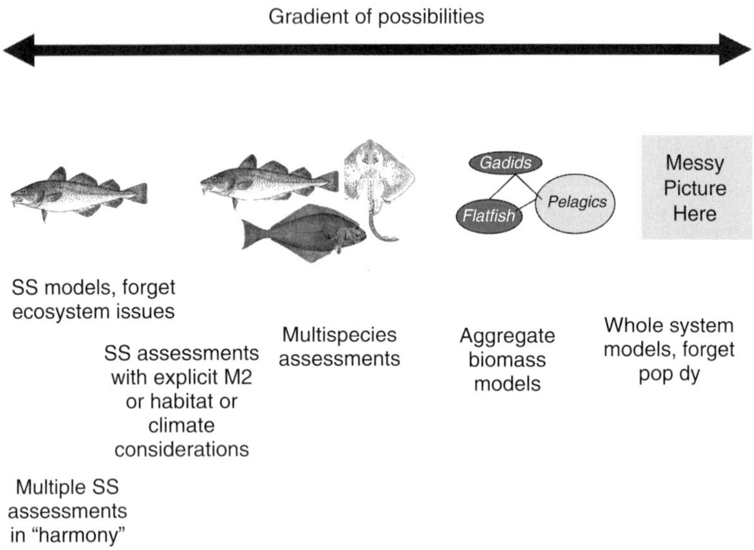

Figure 2.1 A gradient of approaches for ecosystem considerations in fisheries management. (Adapted from Link 2002b.) SS, single species; M2, predation mortality; MS, multispecies; pop dy, population dynamics; Messy Picture Here, your favorite food web horrendogram with additional climatic and oceanographic and market forcing functions overlaid upon it.

accounting for, and including all factors in a holistic, synthetic, integrated fashion. Although mildly distinct from EAF, the two are used interchangeably. I tend to think of EAF as starting with a stock and building up to include as many factors as are appropriate, building from the left of Figure 2.1 and working up the gradient. In contrast, EBFM starts with an ecosystem perspective of fisheries production and addresses specific interactions among species and with the environment, building from the right of Figure 2.1 and working down the gradient. Although fisheries are emphasized in both terms, the inclusion of marine mammals, protected species, non-target species, etc., is usually implied for EAF and is more explicit in EBFM.

There are many stories about why we have all these similar terms. For instance, some state that EBFM does not translate well into pictographic languages like Chinese, but EAF does (sensu Sissenwine and Murawski 2004; M. P. Sissenwine, personal communication, 2000), thus its adoption in international institutes like the United Nations' Food and Agricultural Organization (FAO). Others note that EAF allows for the fishing sector to remain with classical SS approaches or to include enhanced considerations, while preparing to engage with other ocean use sectors

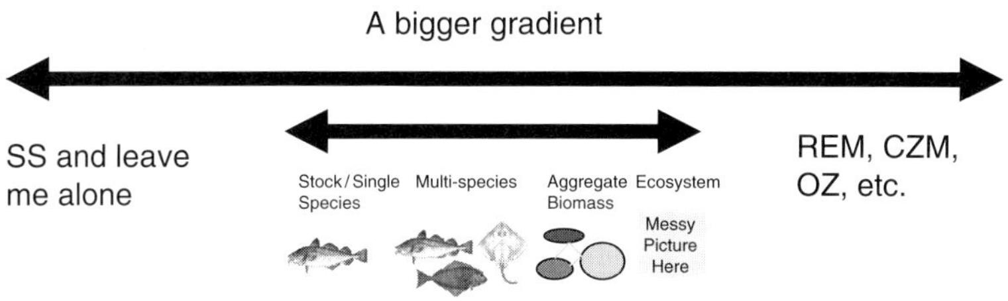

Figure 2.2 A bigger gradient. SS, single sector or single species; middle of graph adapted EBFM from Figure 2.1; REM regional ecosystem management; CZM, coastal zone management; OZ, ocean zoning, etc.; all implying consideration of broader ocean sector use.

(K. J. Sainsbury, personal communication, 2004). The point is that the distinction between approaches within and across sectors (Figure 2.2) is clearly important. The distinction between whether or not management is *ecosystem based* or is using an *ecosystem approach* is likely more trivial. I have tried to follow the emerging convention of using "ecosystem based" to emphasize an entire system perspective first, and "ecosystem approaches" in instances emphasizing a sector or stock first. Although a minor distinction, it may be useful. However, even I also use the specific terms EAF and EBFM interchangeably throughout this work, depending upon from which side of the continuum of approaches (Figure 2.1) I am starting. As the title of this book implies, I tend to favor the EBFM distinction.

This entire discussion raises the duality of the issue. Are we attempting to manage a species and consider the broader ecosystem (EAF), or manage the ecosystem as an entity to account for species of interest (EBFM)? The answer is a confusing but resounding yes to both. There is often a perceived duality of this choice, but it really is a false dichotomy. Ultimately, what EAF/EBFM means is that we are trying to manage fisheries in a broader ecosystem context, be that context cognizant of other ecosystem considerations or/and other ocean-use sectors (Table 2.2). How fisheries impact, and are impacted by, other components of the ecosystem and other ocean-use sectors is one reason why this issue has emerged to be as prominent as it currently is.

Now that I have given you a sense of what EBFM means, we can examine what EBFM does not mean. It does not mean that we have to wait for a regime shift or climate change or global cooling or other factors beyond human control before we take action. It does not mean that we can eradicate otherwise endangered species because they happen to be eating a portion of a targeted fish stock. It does not mean that we ignore all applicable laws or rules or protections for all

Table 2.2. *Interrelationships among fisheries and other factors when considering ecosystem approaches*

	F	OSe	OEB	AEF	CF	OSt
Fisheries (F)	X	X	X	X	–	–
Other sectors (OSe)	X	X	X	X	–	X
Other ecosystem biota (OEB)	X	X	X	–	–	–
Abiotic ecosystem factors (AEF)	X	X	X	X	–	X
Climate factors (CF)	X	X	X	X	X	X
Other stressors (OSt)	X	X	X	X	X	X

The potential for an effect of the row on the column is noted with an X, while – denotes no significant effect. The "Other sectors" includes: mineral extraction, energy production, navigation/transport, tourism, coastal development, medicinal development; all implying their uses and impacts. "Other ecosystem biota" includes: structure and functional dynamic properties of ecosystems; the entire set of biological communities, biodiversity, non-target species, protected species, etc.; i.e. the full set of biota and biotic interactions. "Abiotic ecosystem factors" includes: habitat, currents, thermal regime, water chemistry, etc. "Climate factors" are broad-scale atmospheric and oceanographic forcing. "Other stressors" includes: contaminants, eutrophication, hypoxia, metal or toxin pollution, harmful algal blooms, etc.

ecosystem components for the sole rationale of maximizing fisheries catches. It does not mean that we can maximize yield for a particular stock at the expense of other species (targeted or not). It does not mean that SS reference points or advice can be ignored. It does not mean that we cannot harvest any LMRs whatsoever from an ecosystem. It does not mean that things are so complicated that we cannot take any action at all. And most appropriately, it does not mean that we can continue to put off the hard decisions of allocation, tradeoffs, and similar considerations while invoking such vague platitudinal goals associated with broad ecosystem objectives (such as maintaining biodiversity, restoring ecosystem health, etc.). Sometimes knowing what we cannot, or should not, do is just as helpful in determining what it is we should do.

THE ORIGINS AND HISTORY OF EBFM

Recall from Chapter 1 that 150 years ago smart minds were thinking about these types of issues. The great debates in fisheries science that occurred through the first half of the twentieth century were prescient in their content and the outcomes, although largely settling on SS approaches and fishing as the way to go and the primary factor to consider, respectively. These arguments provide the intellectual background for the discussions occurring about EBFM today. During the debates in the early to mid 1900s, the scientific titans of the day duked it out; namely arguing over the relative importance of the environment versus fishing when considering fluctuations in fish stocks (Smith 1994). Lankester (and later Garstang) and Huxley were arguing over the concepts of fisheries production and exhaustibility even before the turn of the nineteenth century. As these debates developed, Hjort's observations (leading to Hjort's "critical period" hypothesis) and *mare incognitum* were being espoused (sensu Hjort 1914; see also Smith 1994). Hjort's work examined both density-dependent and density-independent considerations and concluded that the evidence was defensible for either perspective given the timing, scale, and issue being examined. The debates after World War I clearly debunked Huxley's (1884) view (see Borley *et al.* 1923; discussion in Smith 1994, pp. 158–162), but raised the question of whether other factors should be considered in addition to fishing pressure. In many respects, there was much thinking about EBFM-related concepts during this period.

Although arguably one of the most famous intellectual discourses associated with fisheries, the history of the Thompson–Burkenroad debate over the causes of Pacific halibut declines, shows that the issue was never truly settled; nor has it been (generally speaking) to this day (Skud 1975; Walters and Collie 1988; Rose 2000). This debate seemed to spur other debates after World War II about what exactly all the factors were that caused a resurgence and then decline of several fish stocks (sensu Hickling 1946; Graham 1948, 1956; Schaefer 1956; Beverton and Holt 1957). This even though it is now, and by many even then, agreed that both environmental (climate, predation, etc.) and fishing considerations can have influences on fish stocks. Yet the resultant observation from this series of events is that fishing is the one aspect that we can nominally control.

In many ways the debate was further settled by the definition of Schaefer's third tier (Schaefer 1956), whereby density-independent and non-fishing factors were effectively ignored (Figure 2.3). This line of thinking can be traced back to Thompson's (1922) argument for limiting the breadth of fishery research programs. This approach was further codified by a student of Thompson's, Sette, in his proposed research program for the California sardine fishery (Sette 1943). Schaefer (1956) extended this approach when he proposed the sequential investigation of three levels of investigation, corresponding to what he termed surplus production, yield per recruit, and multispecies and environment

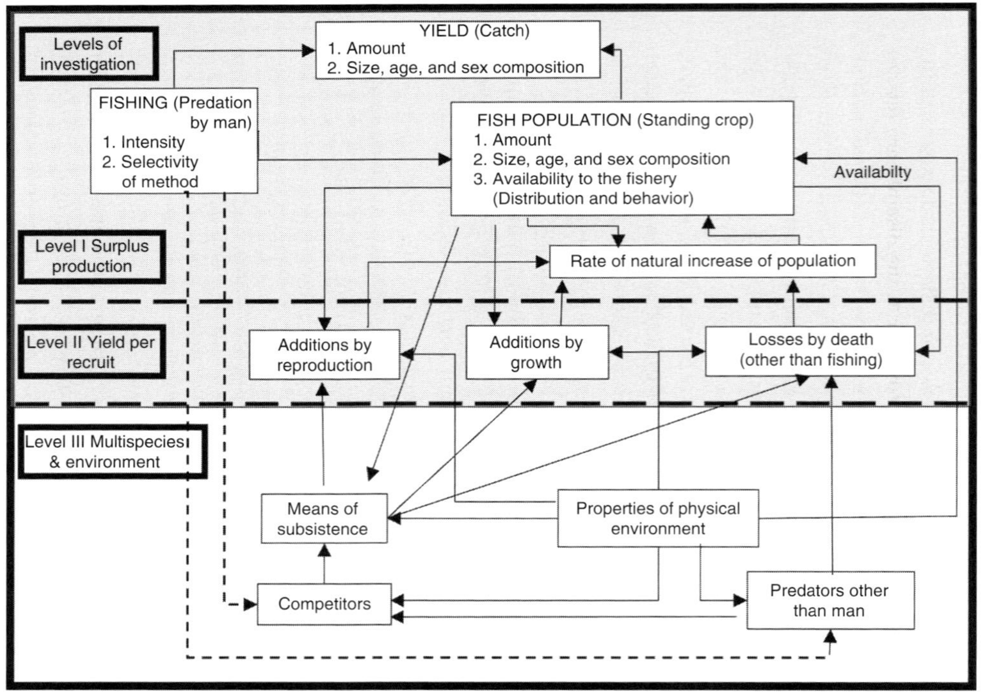

Figure 2.3 Schaefer's three tiers of fisheries investigations. (Modified from Schaefer 1956; via Smith and Link 2005.) The three levels of investigation correspond to what Schaefer termed surplus production, yield per recruit, and multispecies and environment, tiers I, II, and III respectively. The first two levels are shaded gray, indicating the regions of effort that much of fisheries science has focused on.

(Figure 2.3). He argued that the appropriate level of research depended on the significance, and especially the intensity, of the fishery. Gulland (1983) expanded Schaefer's hierarchy in a book that has had widespread influence on training fisheries scientists (Smith 1994), in effect codifying a narrow focus for the field of fisheries science.

The development of Lindeman's (1942) trophodynamic aspect of (aquatic) ecology changed the way freshwater fisheries scientists thought about their stocks. Yet as discussed in Chapter 1, the disciplinary divides were fairly strong at this point, and relatively few marine fisheries scientists were influenced by this integrative precursor to an ecosystem approach. Although a few did consider this approach in marine ecosystems (e.g. Clarke 1946), most marine scientists, particularly those associated with fisheries, were thinking hierarchically rather than integratively.

Although Baird, Sars, Lyman, Lankester, Jordan, Hjort, Gilbert, and many others (Smith 1994) had originally considered a broader view when exploring those factors that could impact LMR populations, Thompson and Schaefer's line of thinking won the day. This emphasis on SS perspectives has carried through to the present for most fisheries endeavors (see Chapter 1), yet the parentage for EBFM-related concepts can be readily traced back to the latter 1800s. Although the specific jargon has changed, the fundamental debate remains.

As these developments were occurring, the International Biological Program and similar international efforts were unfolding from an applied ecology perspective. These approaches used a systems ecology standpoint, in large part building on the Odum brothers' body of work (e.g. 1957, 1969). Here, understanding energy flows and budgets thereof would allow for a better understanding of how (eco)systems worked, and thus allow for improved management of those systems and the component organisms in them that were being utilized.

From there, the sustainable development school of thought emerged. The sustainable development (SD) initiative has also provided a more modern parentage to the EBFM concepts in conjunction with those from 50- to 100-year old fisheries debates. This initiative resulted from an enhanced recognition in the 1970s and 1980s that many of the Earth's ecosystems had been or were undergoing significant impacts from human use. A plethora of conferences and conventions led to the codification of the SD initiative (Table 2.3). The Brundtland Report (WCED 1987), as formally codified in UNCED, led to the formulation of an SD agenda, which can in effect be (under)stated as an effort not to mess up the management of our natural resources today at the expense of future generations' ability to utilize said resources tomorrow

Table 2.3. *Conferences and conventions that served as origins and stimulants leading to the codification of the sustainable development initiative, and subsequent precursors to EBFM*

Stockholm Conference on Human Development (1972)
Convention on International Trade in Endangered Species (1973)
Bonn Convention on Migratory Species of Wild Animals (1979)
UN Convention on the Law of the Sea (1982)
World Commission on Environment and Development (1987)
UN Convention on Environment and Development (1992)[a]
Convention on Biological Diversity (1992)
Jakarta Mandate (1995)
FAO Code of Conduct for Responsible Fisheries (1995)
Reykjavik Conference (2001)
World Summit on Sustainable Development (2002)

[a] UNCED was where the sustainable development agenda was first formally adopted.
See Glossary for further descriptions of these items.

(WCED 1987; FAO 2003). Implicit in the SD initiative is a precautionary approach and the need to consider a broader range of factors, *à la* EBM .

Preceding and concurrent with the development of the SD initiative, there were several literary works that elaborated upon the rationale, necessity, theory, and thinking behind sustainable development and EBM. For example, works by Leopold (1949), Carson (1962), Holling (1978), and Walters (1986) all provided underpinnings for the SD efforts, from which EBM (and EBFM) have emerged. The current concepts, commission reports, and similar items noted at the beginning of this chapter have all generally followed the reasoning (and underlying philosophies) noted in these literary works. The resultant general ideas of EBM have been espoused and applied to other types of ecosystems (Slocombe 1993; Grumbine 1994, 1997; Christensen *et al.* 1996; Haeuber 1996; Larkin 1996; Yaffee 1996, 1999; Yaffee *et al.* 1996; Franklin 1997; Lackey 1998); interestingly, the terrestrial-based concepts of EBM have outpaced those in marine fisheries by at least 10 to 15 years.

SO WHY DO WE WANT TO CONSIDER EBFM?

The justifications and rationales for why we would want to adopt EBFM as an enhancement of or improvement over the current, single stock/species (SS) approaches to fisheries (both management and science) have been noted elsewhere (e.g. Larkin 1996; Botsford *et al.* 1997; NMFS 1999; Link 2002b; Garcia *et al.* 2003). In particular, the benefits of EBFM generally are:

(1) categorically more conservative (precautionary) management recommendations
(2) explicit consideration of non-targeted species, protected species, habitats, etc. (i.e. the ecosystem as a whole) with appropriate precaution built in to the exploitation regimes
(3) directly addressing tradeoffs among and within sectors and also tradeoffs across biomass allocation
(4) the actual potential for simpler management institutions
(5) improved short- and long-term economics for participating fishers
(6) long-term sustainability for intergenerational equity.

Although the debate continues (e.g. Mace 2004; Hilborn 2004; perspective of Eagle in Leslie *et al.* 2008), there is an emerging recognition that EBFM/EAF is necessary.

In simplest terms, EBFM and related EAM may provide a way to avoid negatively fulfilling Graham's Law and also to address the other, multiple-sector societal concerns that are currently difficult to handle by using solely fisheries approaches. There is a clearly recognized need to be holistic, coordinated and integrated in how we approach LMR management.

In sum, what does an ecosystem approach provide that we can't get from SS approaches? As noted in the last chapter, EBFM:

(1) addresses effects of fishing on non-target species, habitat, ecological interactions, and system-wide processes
(2) recognizes that marine ecosystems provide "goods and services" other than fishery harvest
(3) explicitly addresses biomass tradeoffs (in my view the key to the entire issue)
(4) increases leverage from new stakeholders
(5) changes the burden of proof.

It has emerged that even if we used an entirely correct SS approach (as some have argued; e.g. Mace 2004; Hilborn 2004; perspective of Eagle in Leslie *et al.* 2008) there are still many factors that would not be adequately addressed.

Reports from both the Pew Oceans Commission (POC 2004) and US Commission on Ocean Policy (USCOP 2004) have called for changes in how we manage our nation's (and, in effect, the world's) ocean resources. The congruence in both reports is interesting, particularly with a concurrent call for more ecosystem approaches to fisheries. Many initiatives reinforce the relevance and timeliness of EBFM: Canada's Oceans Act; Australia's Oceans Policy; the EU's efforts (as seen by a commitment to the ICES fishery ecosystem plan exercise); some states' (California and Massachusetts) oceans acts or policies; the UN's EAF and related policy changes (as seen in FAO, IOC, UNESCO, and SCOR, among other subagencies); and competing interests in many international resource organizations (e.g. IWC, ICCAT, etc.). In many instances the "whys", "whens" and "wheres" of doing EBFM have been debated and satisfactorily resolved (e.g. Larkin 1996; Botsford *et al.* 1997; NMFS 1999; Link 2002b, 2002c; FAO 2003; Garcia *et al.* 2003; Pikitch *et al.* 2004; Browman and Stergiou 2005; Garcia and Cochrane 2005; Field and Francis 2006; Murawski 2007). Yet the simple fact is that scientific, political, and management institutions remain unprepared when it comes to the "hows" of doing EBFM; implementing EBFM in an operational context has not yet become a widespread reality.

The question remains that if ecosystem processes have the potential to significantly influence fish stocks and have been noted as potentially very important, why aren't these considerations incorporated into fisheries management more broadly and systematically? I submit there are six reasons (sensu Link 2002b):

(1) logistical concerns related to the data availability and collection required to address ecological considerations
(2) limited resources (time, money, and staff) of resource management agencies
(3) concerns that using ecosystem and more holistic approaches will serve as an excuse not to implement clear-cut single species advice
(4) philosophical disagreements in fisheries science about the importance of "lower tier" processes (sensu Schaefer 1956)
(5) the fear that including ecological considerations may increase complexity and thereby add further uncertainty into the management process
(6) finally, the perception that we don't know how to actually do it. Even if we recognize the importance of ecosystem processes and how they influence stock dynamics, if we have abundant and appropriate data, and if we have an institution willing to

support this type of an approach, given the present state of our knowledge, how can we operationally and systematically implement EBFM?

NOW WE'RE TO HOW

There is clearly an obvious need and an increasing interest by the scientific community, resource managers, policy makers, stakeholders, and the general public to address these broader considerations in ocean management, in general, and fisheries management, in particular. Obviously some challenges remain before we can begin to routinely implement EBFM (NMFS 1999; Link 2002b; FAO 2003; Garcia *et al.* 2003). For instance, there are costs of adopting EBFM, particularly relating to data availability, increased complexity, and increased uncertainty, as noted above. However, these concerns and others like them no longer remain viable excuses for not attempting to initiate EBFM. Yet the viability of EBFM and its benefits listed above need to be better documented and presented more clearly and coherently before many of the concerns of the wider interested community will be allayed.

But again, chief of these concerns is lacking a set of approaches on how to make EBFM operational. In large part, the remainder of this book seeks to address just that issue.

SUMMARY POINTS

- There are a lot of terms and acronyms related to ecosystem-based fisheries management; in effect they are shorthand for more holistic approaches to resource management.
- Conceptually, EBFM is not novel; precursor concepts have been extant for well over a century and a half.
- Debates regarding the importance of non-fishing effects on fish stocks have unfortunately polarized the issue into an either/or perspective, when it is understood that multiple factors can influence fish stocks.
- There is no shortage of benefits in implementing EBFM.
- There is no shortage of instances when EBFM implementation is appropriate.
- The challenge remains as to how to implement EBFM.

3

Being audacious

> Be fruitful and multiply, and fill the earth, and *subdue* it; and *rule* over the fish of the sea and over the birds of the sky and over every living thing that moves on the earth.
>
> Genesis 1:28

STEWARDSHIP

The monotheistic basis for, and particularly the Judeo-Christian ethic of, stewardship is pervasive in natural resource management. There are implications from this ethic, and it would be useful to examine them, for they underlie many of the philosophies and considerations for how and why we conduct fisheries management.

The Hebrew word *radah* is used in the commonly quoted and commonly misunderstood passage from Genesis noted above. This word also implies concepts of ruling, having dominion over, taking charge, and being responsible for. Commonly misunderstood, *radah* really means that humankind was given the authority to subdue natural resources for beneficial management, meaning beneficial to both humans and the resources. It also means that humankind is responsible for using these natural resources well. But most telling is that this signifies that humankind is ultimately accountable for how we use those resources. There is no room in the interpretation of this word for pillaging, ruling with an iron fist, pollution, overutilization, or general mismanagement that often is misconstrued by the translations "rule" or "dominion over."

Ultimately, the concept espoused here, and confirmed repeatedly in the Judeo-Christian religious texts, is of stewardship. Stewardship implies a sacred trust, inherently common in many of the indigenous tribes and cultures across the planet. But there is more. Stewardship implies ownership (in the Judeo-Christian sense, by God) but not usually by the steward.

Think of a bank account managed by an account custodian. That custodian is responsible for the performance of that account and is quite typically legally bound to report on the account's activities. There is also an understanding by the account owner, the bank, and the account custodian that the custodian cannot deplete or take any action on the account solely for his or her own benefit. The interests of the owner are paramount, even if the owner is providing for the account custodian with a percentage of the account's earnings. This imperfect analogy helps one to grasp the concept of stewardship; it requires ownership (in a fisheries context, by government, tribes, quota holders, future generations, etc.), but that ownership is rarely acquired by those entrusted to manage the resources.

More to the point, stewardship implies management. Stewards are accountable for how well they manage the resources entrusted to them. Stewards who manage resources wisely are routinely lauded in the religious texts. Those that do not are presented as examples of shame. Thus, as a collective group, humankind has a stake in how well we manage our natural resources. In the Judeo-Christian and in many indigenous cultural contexts, collectively we will be evaluated on our performance of how well we have managed the natural resources of our time.

Finally, explicit in this concept of stewardship is a mandate. There is an imperative statement that humankind will husband the natural world to our own benefit, yes, but also to the benefit of that world. There is no room given in this mandate for us to do so if we feel like it or if it is convenient. Rather, there is no other option. The mandate implies that we must take care of and manage the natural resources of the world. There are numerous initiatives, journals, and fields of study on "creation care," environmental stewardship, and environmental ethos (e.g. www.creationcare.org; www.arocha.org/int-en/index.html; www.earthcareonline.org; www.earthministry.org; www.asa3.org/ASA/topics/environment/index.html) that delve into this topic much more deeply than the treatment I have provided here, but even a cursory presentation of these concepts is warranted.

Why is it important to intentionally identify the philosophy behind the concept of stewardship? Without this underlying ethic and its explicit mandate we would not have a culture of managing resources. Additionally, before there was enabling legislation that required humankind to manage natural resources, it was understood in most cultures, not solely those influenced by this Judeo-Christian ethic, that caring for the Earth and all that is in it was a divinely appointed task. That is, it is a God-given duty for us to be wise stewards of the natural resources on this planet. And it is sobering to recognize that the implied provisions ordained for our survival are linked to our prudent management of these natural resources.

PRIORITIES

Another important consideration based upon the Judeo-Christian ethic of stewardship is that it gives us a list of priorities. In the quote above, the concept of *subduing* the Earth is intriguing. Yes, this obviously says that humankind is to subjugate the world's natural resources. It does not mean that such subjugation should devalue the natural resources. But we first need a proper valuation of natural resources to understand how best to utilize them.

Most cultures and belief systems, particularly the monotheistic systems, have a "natural order of things" whereby inanimate materials are subjugated to plants, which are subjugated to animals, which are subjugated to humans (e.g. see Psalm 8). This is a clear list of ordering (and prioritizing) the values we place on different kinds of life. For instance, the earlier parts of the Genesis story (Genesis 1–2) list land (habitat), water and light (nutrients), plants (lower trophic levels), animals (upper trophic levels), and then humans. Of course, not all belief systems concur with this valuation, but most recognize this perceived "pecking order." It is also true that not all belief systems that do concur with this valuation always implement it in practice. Yet, for the most part, having such a list of general priorities can be helpful for the difficult task of decision making, conflict resolution, and addressing tradeoffs in natural resource management. Additionally, at some level all belief systems recognize four main features of natural resources: (1) the interconnectedness among living things; (2) the interconnectedness between living and nonliving things; (3) the hierarchical interdependence of living organisms (think Eltonian pyramids); and (4) the cyclicity of human bodies returning to inanimate, elemental composition upon death. This general consensus of inherent philosophies can be built upon.

Admittedly, I have simplified this entire topic; for further reviews and discussion on the interplay between religion and ecology see Callicott (1994); Waskow (1995); Berkes and Folke (1998); Berkes (1999); Ludwig *et al.* (2001).

Notwithstanding the interplay between religion and ecology (again, for further reviews and discussion see Callicott 1994; Waskow 1995; Berkes and Folke 1998; Berkes 1999; Ludwig *et al.* 2001), it is important to recognize the philosophical underpinnings of our modern worldviews of natural resources. Several of the differing philosophies opposed to the Judeo-Christian ethic of stewardship espouse equality of all living things, in some sense accurately recognizing the interconnectedness and interdependence among life, as previously noted. Yet, upon further inspection, these philosophies can be rejected as a basis for natural resource stewardship for at least four reasons. First, they devalue human life, which, regardless of belief system, is entirely reprehensible. Second, such philosophies did not necessarily lead to the development of the stewardship ethic upon which natural resource management is based. Third, extended to their logical conclusion, such philosophies run counter to the generally recognized pecking order of species value. Finally, such philosophies do not always recognize that collectively humans have the responsibility for stewardship, a responsibility not placed upon plants or animals. There is no doubt that humans have unique capabilities that have allowed us to manage, or mismanage as the case may be, the world's natural resources such that the scale and scope of our impacts have become astounding. No other species that are capable of engineering landscapes or utilizing simple tools (e.g. beavers, other primates) are able to purposefully and directly influence the entire biosphere. For instance, how audacious is it that at a global scale we have driven species to extinction, drained wetlands from large portions of entire continents, and are overfishing the vast majority of the world's fish stocks? Thus, a recognition of underlying philosophies helps us to communicate with all stakeholders engaged in natural resource management and also provides the basis for the stewardship ethic generally understood in natural resource management.

There are additional benefits to stating our philosophical underpinnings explicitly: it can be of practical use, provide the basis for common consensus, and lead to development of clearer management objectives. The concept of subduing the Earth for the purpose of wise stewardship can provide a basis for prioritization. A clear list of general objectives could then follow.

In this sense, it is recognized that plant life (or more precisely and generally, primary producers) is needed to produce animal life (i.e. secondary consumers), but is typically not valued as highly as most animals are. This is not to say that grains and greens and trees and

Table 3.1. *Hierarchical prioritization of overarching EBFM and general resource management objectives based upon the Judeo-Christian stewardship ethos*

Level	Objectives	Supporting principles and concepts
Top	Human existence	Human rights
	Other species existence	Biodiversity, endangered species
	Individual stock/population health	Sustainable use
	Persistence of particular human communities	Cultural values
	Ability of other people to make money	Fairness in competition
Bottom	Maximizing individual profit	Individual pursuits

lumber are not important, rather they typically do not command as much value per unit mass in the marketplace compared with organisms higher up the food chain. As an example, go to your local grocery store and compare prices per unit mass of material in the produce aisle versus material in the meat department. With sentimentalities and other philosophies aside, those believing in this implied pecking order would then state that no individual of any other species, nor any natural resource (e.g. oil) is as valuable as any individual human life. Conversely, the logical extension of some philosophies is that the inherent evilness of humankind is such that the planet would be better off without us. Granted, humans are capable of incredible iniquity, but taking the middle road, the stewardship ethic recognizes that humans do indeed have a place and a role on the Earth.

How would this prioritization play out when we attempt to manage fisheries? Particularly given that fish stocks are a shared resource in most areas, I would argue that it might lead to the following (Table 3.1). Economic objectives shall be subordinate to social objectives, which shall be subordinate to biological objectives, which shall be subordinate to human existence. That is, maximizing individual profit, not in itself necessarily a bad thing, should not override the ability of other people to earn a living nor of communities or cultures to persist, which should not override the health of a resource stock, which should not override the extinction of species, which should not override basic human rights (starvation and survival, as in some developing world situations). Audacious to actually state these positions, isn't it?

Granted, statements like these can be difficult to accept. In many ways this hierarchy runs counter to the culture of capitalism and free market economies found in the Western world. In many ways it runs counter to the philosophics and worldviews that tolerate miserable human conditions. In many ways it places the persistence of supporting biota above any particular human culture or community, a view apt to provoke heated debate. Yet it also recognizes the need for some form of control on collective human endeavors, with clear prioritization on the value of all human life, or else other species, and in some cases other humans, will suffer tremendously. Perhaps this list has not been ordered entirely correctly, but it should be close enough to spark debate of how best to modify it and it should be close enough to encompass the underlying stewardship ethic. Other than general suggestions (e.g. Brundtland Report, WCED 1987; UNCED; NMFS 1999; FAO 2003), I am unaware of a list like this stating what the general prioritization of fisheries management objectives should be, such that they are applicable for the entire human condition.

The irony of our times is that we have inverted the prioritization hierarchy and then wonder why natural resources are degraded. We attempt to make the lowest priority on the list our highest without realizing that we need the other considerations first. That is, we can't have something we value highly (say, copious money) without having the foundational items first. I personally don't mind, even support, and have certainly benefited from the capitalistic system; yet not to the point of putting the proverbial cart before the horse. If basic human needs and rights (i.e. Maslow's pyramid) are not provided for, individuals will not care about their immediate environment. By placing the persistence of a species or the value of other natural resources above human life, we downgrade individual people such that they lose the incentive to care about how well they should steward those resources. Valuing a protected species above another human being, an individual human life, ultimately will serve neither one of them. Furthermore, without a well-managed, healthy set of living natural resources, there is no basis for ecologically supporting and sustaining human communities and cultures. Impediments or disruptions to routine ecosystem services and functions make it difficult for societies to maintain and persist for several obvious reasons. Obtaining wealth while disregarding ethical, legal, or moral considerations is not a good way to stay in business for the long term. One can disregard societal rules and regulations for only so long before the rest of society notices, usually to the point of irritation and resulting in corrective action.

Attempting to make money without stabilizing those underlying factors (biological production basis, social fabrics, etc.) that are required to support such endeavors seems myopic and almost always leads to duress if not extreme conflict. No, there is nothing wrong with making money or even lots of it; rather there seems to be a disjunct in doing so while ignoring the basis for *being able* to do so. That is, in many ways we have begun to kill the proverbial goose that lays the golden eggs.

MARINE CAPTURE FISHERIES ARE NOT FARMING

The concept of stewardship implies that we can actively manage a resource with the goal of wise utilization. In some cases, heightened levels of protection or remediation are needed for degraded resources, but there is an understanding in the concept of stewardship that the stewards are able to utilize resources as they manage them. But there is less recognition that we need to be stewards for both cultivated and wild organisms. Too often, our thinking and ingrained management customs assume farming as the default perspective for natural resources.

Much of the stewardship school of thought for natural resources arose around agricultural or pastoral situations. Understandably many of the living resource management principles are based on these systems, where for centuries, cultivation and husbandry produced a more stable source of food and income than hunter-gathering. Even the more nomadic hunter-gatherer societies still emphasized some husbandry and cultivation, albeit usually in less structured ways (Diamond 2005).

Think about common fisheries terms used today. We have stocks, be they spawning (brood) stocks or standing stocks or stock biomass or spawning stock biomass. We have stock–recruit relationships. We have harvests. We have harvest rules, yields, yield per recruits, spawning (breeding) grounds, spawner-per-recruit relationships, and spawning seasons. We have feeding grounds. We have feeding seasons. We have home ranges. We have production, especially surplus production. We have culling. And so on. Where do these terms come from? Definitively from a farming context. Yet is a farming context and framework appropriate for marine capture fisheries? Maybe in some sense for aquaculture, but even accounting for that, the vast majority of the fishes marketed and caught today are, in effect, wild animals (FAO 2007).

There are several assumptions about farming that do not hold for marine ecosystems and the fisheries in them. One is that farming (or by extension, forestry) is an areal-based activity. Farms are clearly defined, managed, and operated in distinct units of land. Until recently, spatial management was not a widely used fisheries management tool, largely because of the difficulty of transposing the farming analogue to continental shelves or the high seas, given the oceanic nature of these ecosystems and their main driving processes.

The second main feature about farming that does not clearly translate into marine ecosystems and resources is one of obvious ownership. In a farming context, an individual buys the land or rents the right to use it. In marine fisheries, who owns the fishes? In most cases this is nominally the government, but situations whereby fishers pay for the right to access fisheries are the exception, not the rule (sensu Ostrom *et al.* 1999; Charles 2001; Edwards 2003). It has been noted that a large contributing factor to the fisheries problem is the tragedy of the commons (Hardin 1968), whereby everyone has access to the fish and in that sense owns them. Such a situation with vague or amorphous ownership produces more conflicts in marine fishing than in farming.

Additionally, farms and forests do not move. The ocean does, and the mobility of inhabiting organisms also distinguishes farming or forestry from fishing. Domesticated animals are kept in much more confined ranges than are wild animals. Even wild terrestrial animals have much smaller ranges than marine fishes (Steele 1985, 1991; Link *et al.* 2005c). For instance, managing one 500-kg animal, say a cow or ox, on land means supplying enough food, water, and waste removal along with some space for this animal to move about perhaps tens to maybe hundreds of acres. Now let us say we manage one 500-kg animal, perhaps a tuna or whale, in the ocean. This effectively means we have to track that animal throughout an entire ocean basin. In other words, although we don't need to directly supply the bioenergetic needs for this animal, we need to evaluate, monitor, and either harvest or avoid harvesting this animal over a significant fraction of the Earth's area. The same scaling often holds for even a 10-kg fish. The resource plots for farms or forests are usually much smaller than individual states, provinces, or counties. Conversely, resource plots in the ocean are vast (e.g. Figure 3.1). The spatial scale and scope of organisms in the ocean is orders of magnitude greater than those required for farming or forestry. The extent of operations is similarly much larger in scale for fishing than farming, with the obvious distinction of scale following the mobility of the resource "patch" and its associated biota.

There is directed, intentional "natural" selection on farms; there is directed, unintentional natural selection in fisheries. A major distinction between organism genetic selection on land and that in the oceans is that there is direct husbandry for specific traits in the first case.

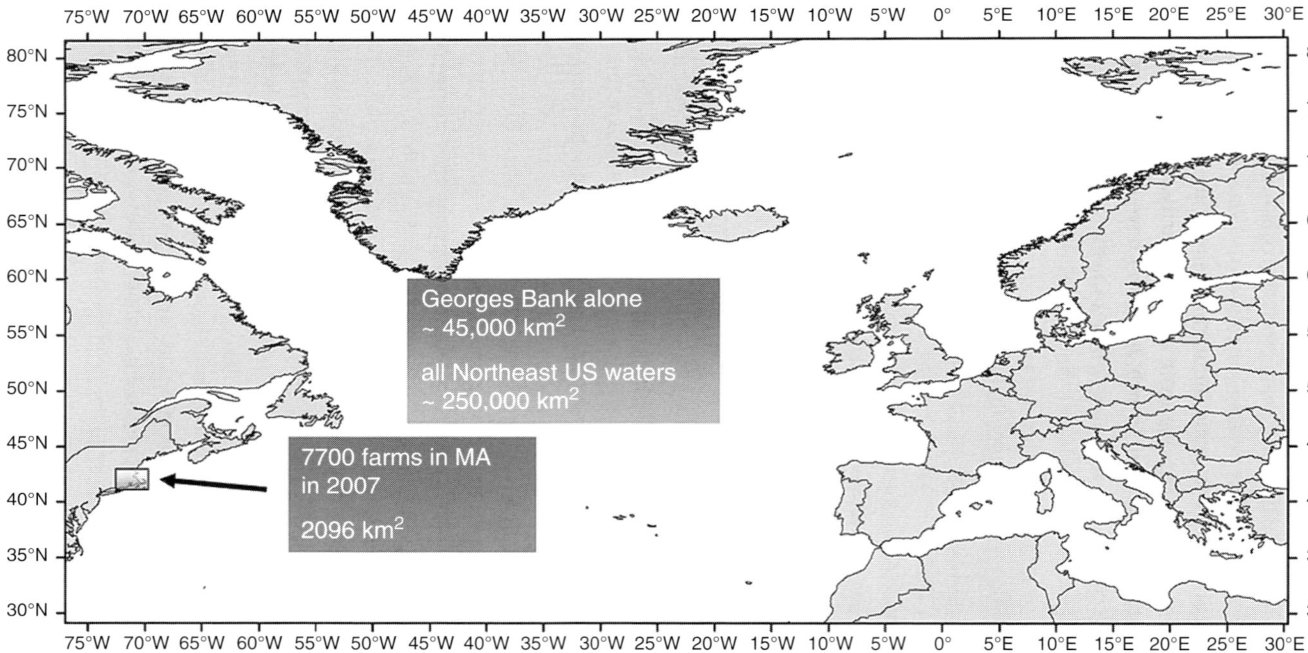

Figure 3.1 Scale of marine capture fisheries in one ecosystem relative to farming in one US state. (Data for Massachusetts farms from US Department of Agriculture, 2009.)

Farmers typically wish to grow more-faster-stronger-bigger-healthier-disease resistant-whatever, in short, *better* organisms for their own benefit. After several centuries of practice, humans have effectively selected those genotypes to maximize what a farmer or forester would like. Conversely, any selection pressures occurring on marine fishes may not be desirable (e.g. smaller, faster-growing, younger age-at-maturity, less fecund, less viable eggs, etc.; Conover and Munch 2002) and have certainly not been intentional. Any selection on marine fishes may in fact not be assisting fishers to achieve maximal benefits from the organisms they are inadvertently husbanding. This distinction between farming and fishing is not trivial in that it contrasts deliberate, intentional choices to maximize production on the one hand versus accidental, unintentional outcomes that serve to minimize production on the other.

Additionally, the largest individual organism biomasses on farms or in forests are at the low trophic levels. The opposite is true for the oceans, with the largest individual organism biomasses at upper trophic levels. Thus, in these marine ecosystems it is more energetically inefficient to make upper-trophic-level fishes like tuna or iconic species like whales than it is to make a cow or white pine. This directly relates back to the distinctions in spatial scale required to produce a terrestrial and marine organism of approximately the same individual biomass. But it also belies the thinking that we can simply set up a feed lot or a plot of land and expect to get the same response in marine organisms as we do in terrestrial ones.

Finally, on land, multiple uses are typically managed with obvious clarity. That is not so in the ocean, with multiple uses often in conflict. I do not mean to imply that there is no conflict in land usage nor that the oceans have had no multiple-use sector planning. Rather, walking, bird watching, deer hunting, and camping in a large forest that also serves as a watershed buffer is much more commonly accepted and done than is attempting to accommodate multiple ocean-use sectors or even different fleets targeting the same fish stock (directly or via bycatch) or LMR in a given area of the ocean.

Thus, given the distinction between farms or forests and fishing in the ocean, it would be wise to be aware of, consider, and perhaps even challenge the underlying conceptual philosophies for resources management that have been based on terrestrial systems. As shown in Figure 3.2 (FAO 2007), one of the key similarities between fishing and farming is that they both provide food for human consumption. The amount consumed varies by country and ecosystem and species, but it can be a significant amount of the supply for human food.

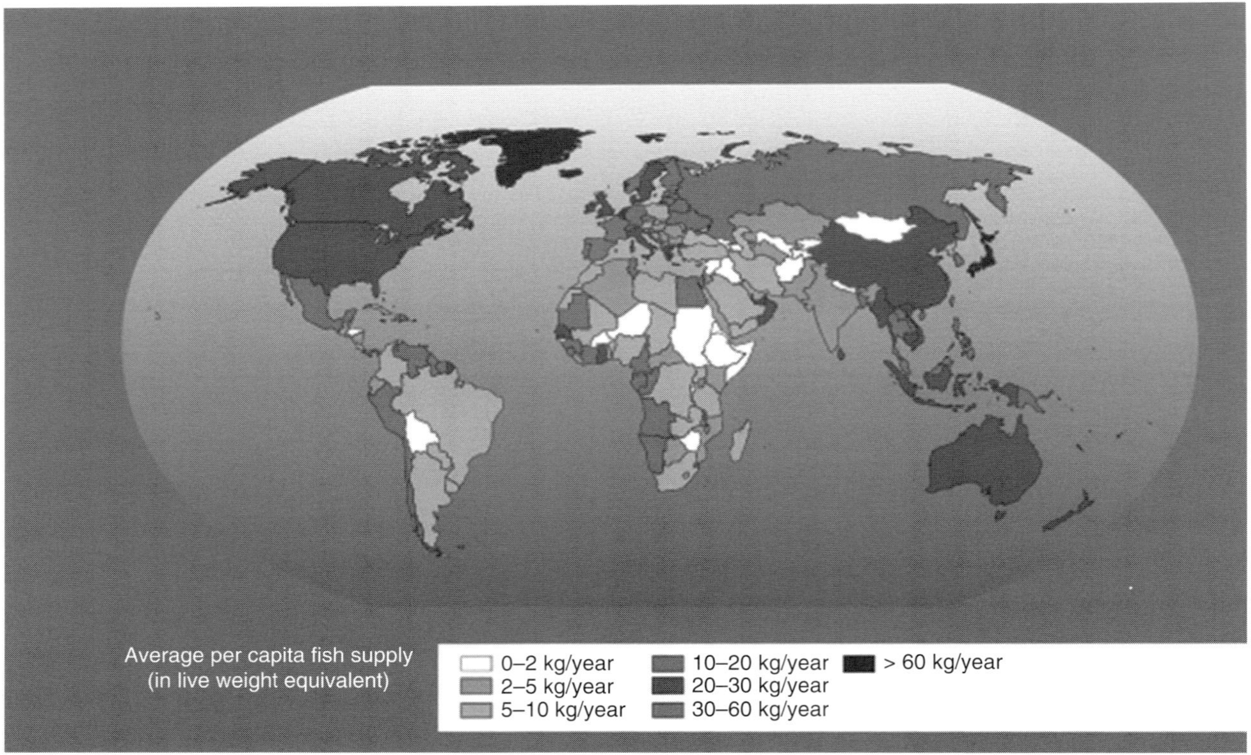

Figure 3.2 A global perspective on the supply of fish as food, per capita, average 2001–2003. (Adapted from FAO 2007.) This figure shows one principal similarity between farming and fishing: the importance of the resultant products from each as food.

AUDACITY

How audacious is it that we attempt to manage free-ranging, wild-living species over such broad spatiotemporal scales? Surely it is audacious that we realize the stewardship humans have been entrusted with to manage natural resources. Surely it is audacious to attempt to prioritize our objectives in doing so. But it is incredibly audacious that we attempt to manage species that encompass significant fractions of the Earth's surface. Yet in some ways, these audacious points pale in comparison with the realities of contemporary fisheries.

There has been a change in perception regarding fisheries over the past century. In the 1880s, Huxley (1884) believed that "probably all the great sea-fisheries are inexhaustible," and in some places that view has persisted to this day. The thinking then, and by many today, was that the oceans were so vast and the organisms so fecund, how could humans possibly impact them? Yet as noted in Chapter 1, we are indeed having an impact on the majority of the world's fish stocks, with over 70% of them at fully exploited or overexploited levels (Figure 1.1) and global landings of marine capture fisheries being effectively stable for the past 20 years (FAO 2007). Additionally, we are actually impacting other features of the world's oceans via these fisheries activities (see Chapter 1).

When you stop to think about it, what we are doing is quite remarkable. Impacting fish stocks on a global scale is indeed audacious. As such, our thinking on how we can improve the global fisheries situation, so that we can be better stewards of those resources, needs to be equally audacious. In the following chapters we will explore some ecosystem-based approaches and concepts that, relative to currently applied fisheries scientific and management methods, may seem a bit audacious. But my argument is that those improvements are very minor propositions relative to the audacity of the impacts we have already had on the world's ocean resources.

SUMMARY POINTS

- Collectively, humans are responsible for the stewardship of the natural resources on the planet.
- Fishing is not farming, but much of our underlying thinking about it belies this assumption.
- It may seem audacious that we are responsible as stewards, that fishing operates at such vast scales, and that we would then try to do EBFM.
- But most audacious of all is that we have impacted fishery systems globally to the scale and extent that we have.

4

Framework for scientific information to support EBFM

Every truth passes through three stages before it is recognized. In the first place it is ridiculed. In the second it is violently opposed. In the third it is regarded as self-evident.

Arthur Schopenhauer (attributed)

A TRIAGE LIST

It is not always appropriate to consider every ecosystem factor in living marine resource (LMR) management. So when is it appropriate, or even imperative, to consider the broader factors? It would be advantageous to have a set of criteria to use in making the decision to embark upon EBFM. A checklist of things to consider would be helpful, as would a set of examples of when EBFM has been shown to be important.

The case studies presented in Chapter 5 serve as generalizable examples for when the implementation of EBFM is needed. But more specifically, some of the characteristics noted in Table 4.1 can help delineate the major features of any particular ecosystem and set of species warranting an ecosystem approach. Forcing oneself to compile information on the geography, physics, biota, ecology, and fisheries for any given ecosystem is a valuable exercise in itself. Evaluating these factors is useful to determine the relative importance of trophic interactions, abiotic factors, fishing mortality, and so forth. No standard set of conditions and rules applies to all ecosystems, and attempting to prescribe a common set of goal functions for each ecosystem is imprudent. I will not advocate any particular methodology, model, or approach to address this list of issues, as their importance changes from ecosystem to ecosystem. Conversely, a generic list of issues should be considered when seeking to implement EBFM so as not to overlook

Table 4.1. *A general list of issues that need to be considered for implementing ecosystem considerations into fisheries management*

Geography of the ecosystem

- What are the key features of the ecosystem under consideration? For example, is the system relatively open (e.g. mid sea, continental shelf) or closed (e.g. river, small lake, bay)?
- How big is the ecosystem?
- What are the important, dominant, and unusual physiochemical factors in a system?
- Is there a prominent geographic, bathymetric, or similar feature that defines and dominates the system?
- What are the political boundaries and jurisdictions that govern the resources in a system?
- How dense is the human population in or near the ecosystem?

Key species

- What are the key species in the ecosystem? Certainly a list of commercially exploited species is required, but noncommercial yet ecologically valuable species should also be included.
- What are the key attributes of these species?
- Are the species involved relatively slow growing with a long life span, or are they more *r*-selected?
- What is the size of the species in the system?
- What is the extent or range of the species of interest in a system? How are they spatially distributed?
- How economically valuable are the exploitable species?
- Are there any keystone species? Are there any dominant species?
- What is the functional role of the key species?
- Have the life history parameters for a species changed over time (e.g. faster growth, earlier age-at-maturity, etc.)? Have they even been determined?
- Are there any species particularly susceptible to an ecological process?
- Are there any specialists?
- Are there any species that are near extinction?
- Are there any species that have an excessively high linkage density (high number of predators or competitors)?
- Are there any species that have sensitive or low-output reproduction?

Abiotic factors

- Are there certain spawning or nursery grounds that merit protection?
- Is there a particular habitat feature (e.g. stacked cobble, sea grass, or oyster beds) that enhance the survivability of juvenile fish?
- Is there a particular area that is optimal for growth?
- Are there particular features such as a thermocline or frontal boundary that aggregate prey for fish feeding?

Table 4.1. (cont.)

- Has the habitat been altered in any way?
- Are there any toxins in the system that can kill or chronically impair a species?
- Is the system susceptible to large-scale perturbations such as a hurricane?
- Is there the possibility that a hypoxic zone could develop?
- Are other forms of pollution prevalent and significant?
- Could harmful algal blooms develop, and if so, what effect would they have on key species?
- Is there evidence of a long-term regime shift in temperature, salinity, atmospheric pressure, upwelling, or similar meteorological factors? Have circulation and current patterns changed across time?
- How strong are tidal influences? Have they changed?
- Are certain life stages or certain species particularly susceptible to environmental change?

Species interactions

- Have the interactions between species been identified? If so, can they be quantified?
- What is the amount of food required to maintain a predator population at a certain size structure and abundance?
- What is the total number of individuals removed by all predators of a particular species?
- Are the interactions between species strong and tightly coupled, or is it a system of generalists with weak species interactions?
- Is there one species that is clearly a competitive dominant?
- Is there evidence of dietary, spatial, or other resource overlap?
- Is there an indication that resources may be limiting?
- What are the key resources in a system for fish, plankton, benthos, etc.?
- Is there a potential for conflict among fisheries targeting different species?
- Are there management protocols in place to objectively resolve these conflicts?

Aggregate properties

- What is the productivity of the ecosystem? Has it changed across the life span of key species? How does this affect carrying capacity for upper trophic levels?
- Similarly, have there been changes in secondary production in the system?
- Is an understanding of the dynamics of lower trophic levels such as benthos or zooplankton essential for the key fisheries?
- Is the food web tightly connected to the nutrient dynamics of a system such as an estuary or small lake?
- Are there significant guilds in the system?
- How is the energy and biomass of the ecosystem partitioned amongst different functional or aggregate groups?

Table 4.1. (cont.)

- What is the dominant group?
- Has this group remained dominant across time? If not, what caused the changes?

System-level properties

- Are there other ecosystem goods and services that compete with a fishery or a particular species? Conversely, are there synergisms between different user sectors?
- How does a fishery interact with other sectors that use an ecosystem (e.g. tourism)?
- Are there protocols to address these potential conflicts or encourage possible collaborations?

The fisheries context

- What type of fisheries have been in the system (commercial, recreational, artisanal, etc.)?
- What type of gear has been and is being used?
- What is the historical level of fishing effort on key species in the system?
- What is the current level of fishing on key species in the system? How does this influence non-target species, trophic structure, habitat, etc.?
- What are current landings and discards?
- Can we adapt gear or else choose to target species as a group that have high technological interactions?
- Where are stocks relative to historical levels of abundance (declining, collapsed, or recovering)?

Source: Adapted from Link (2002b).

a dominant but understudied (or disciplinarily tangential) process (Table 4.1). So, on the one hand, each ecosystem is admittedly unique, yet, on the other, there is a need for a standardized set of processes and factors to be considered.

Thus, how can one use the issue-specific answers to questions posed in Table 4.1 to initiate EBFM? The primary answer is triage. Determining which process is the most important in an ecosystem is valuable and should not be discounted, as is knowing the magnitude and relative importance of the major processes in an ecosystem. Second, answers to these questions could be used to modify existing fisheries management advice in a quantitative way. For example, the International Commission for the Northwest Atlantic Fisheries (ICNAF) used two-tier quotas that set limits on single-species as well as aggregate biomass removals from the northwest Atlantic (Brown *et al.* 1976). In this vein, one could set aside a prescribed amount of a harvested forage fish to ensure adequate food for protected species or predators that are also commercially harvested. I will return to this theme in later chapters. Finally, answers to these questions could be used to qualitatively modify the direction of LMR management advice. For example, if an environmental regime shift has occurred or available habitat has increased, one would expect that carrying capacity or production rates for some species could be modified (up or down) based upon the new conditions.

To initiate EBFM, we do not need to answer every one of the questions listed in Table 4.1. We do not need to ever have a full and complete understanding of ecosystem processes. However, we do need an investment in the data required to elucidate the magnitude of ecosystem processes. We do need to expand multispecies, trophodynamic, biophysical, and ecosystem modeling and monitoring, and we do need an explicit recognition that ecosystem processes can have significant influence on fish stocks, ultimately to the point of being formally addressed.

The Ecosystems Principles Advisory Panel (NMFS 1999), Pew Oceans Commission (POC 2004), US Oceans Commission (USCOP 2004), and similar such entities have all advocated inclusion of ecosystem processes into more holistic fishery management. I submit that the questions listed in Table 4.1 should serve as a useful start to that end. I also assert that addressing even a few of these issues is feasible and can improve our management advice and scientific understanding. Given the increasingly multiuser interests, allocation demands, and shrinking amounts of fisheries resources, we need to begin formulating advice that incorporates these ecosystem processes, now! In many respects, we already are (see Chapter 5).

The challenge remains to quantify the magnitude of ecosystem processes relative to fishing. As noted in Chapter 2, fisheries managers and scientists have long appreciated the importance of addressing this challenge, but recognizing the consequences of ignoring ecosystem considerations and developing a feasible set of widely used tools to implement these considerations has been lagging.

DON'T THROW OUT THE BABY WITH THE BATHWATER, BUT DO CHANGE THE WATER ONCE IN A WHILE

Given the limitations of what single species (SS) approaches cannot, and were not, designed to address (see Chapter 1), they still very much have merit. The advice produced from SS approaches is not broken; it is simply not often utilized. In fact, many of the models, data, and approaches remain germane for targeted species (TS), non-target species (NTS), and protected, endangered, or threatened species (PETS). Furthermore, we should not ignore the information from these SS approaches. The scientific advice is generally sound in these approaches; it is just often not acted upon for a variety of social, economic, or political reasons (Rosenberg *et al.* 1993; Beverton 1998; Caddy 1999; Caddy and Cochrane 2001; Sissenwine and Mace 2003; Caddy and Agnew 2004). Several reports assert that proper SS management will go a long way towards EBFM (Beverton 1998; NMFS 1999; Sissenwine and Mace 2003; Hilborn 2004; Mace 2004; see perspective of Eagle in Leslie *et al.* 2008), and in many respects that is true. Even if EBFM is fully implemented, we will still need to know the status of TS and PETS stocks.

Given the importance of SS approaches, I humbly submit that we must broaden our assessments. It is recognized that doing SS perfectly is still not all that EBFM entails, nor will it fix all the problems associated with the global fisheries crisis (see Chapter 1). Because EBFM needs to consider a whole host of other processes and factors (Larkin 1996; Link 2002b; Garcia *et al.* 2003), staying at one end of the gradient (Figure 2.1) is not an option.

Fortunately, there are now ways that we can expand our fisheries science approaches to better address the other issues associated with EBFM. Beyond the classical SS models, there are SS models with additions that account for ecosystem processes (e.g. adding in M2, habitat-meditated changes to K, temperature-toggled B_{MSY}, etc.), multispecies models that specifically address species interactions (e.g. MSVPA, MSProd, MSFOR, etc.), and aggregate biomass analogues of SS models (e.g. $F_{0.1}$ for all flatfish, B_{MSY} for trophic level IV, etc.), all reviewed in Hollowed *et al.* (2000a), Whipple *et al.* (2000), and Plagányi (2007). Finally, there are system-level reference points, directions, and surfaces with associated control rules (Link *et al.* 2002a; Collie *et al.* 2003; Link 2005). Sainsbury *et al.* (2000), Witherell *et al.* (2000), Constable (2001), and Hall and Mainprize (2004) provide pragmatic examples of instances where these broader ecosystem factors are being considered in a LMR management context.

This combination of possible modeling efforts reflects the gradient of approaches for doing EBFM (Figure 2.1). In later chapters, I will explore each of these in further detail. The point is that the key for successful implementation of EBFM is to expand efforts and use at all points along the gradient, which includes classical SS approaches (albeit perhaps modified, or doing them less frequently in a fuller systems context).

LEVELS AND USES OF MANAGEMENT ADVICE IN EBFM

Given the inherent complexity of ecosystems, it is useful to minimize complexity when discussing EAM. To that end EBFM can be split into three elements which are often used interchangeably and incorrectly, which leads to further confusion in an already complex topic. For natural resource management in general, there are three levels of scientific advice that can be provided: heurism, strategic management advice, and tactical management advice.

Heurism in an EAM context is important. How well do we understand a specific ecosystem? What are the key processes and components of a system? How does a particular ecosystem function? What factors are dominant and what is the order of magnitude of fishing relative to other considerations that can structure an ecosystem? Answers to these and similar questions (i.e. the triage list in Table 4.1) are imperative for meaningful EBFM. This might even represent some qualitative recommendations (e.g. directional) as noted above. The majority of the work on EBFM has probably been in this area. However, providing enhanced insight into the understanding of an ecosystem is not the same as providing management advice in a decision-theoretic format. Certainly background information should play into ecosystem-based decisions, but that information alone is insufficient to serve as control rules and reference points, directions, or surfaces.

Strategic management is a critical element of natural resource management in general. It is amenable to, and I submit even imperative for, EBFM. This level of management advice is not setting specific quotas or targets per se; rather it attempts to provide bounding conditions for the overall system. That is, the scope of this level of management is bounding of all the particular policy objectives that are feasible for an ecosystem. For instance, from satellite imagery we can estimate the primary productivity of a particular ecosystem as x units of biomass per year per unit area, or that over history the total landings of all species in a system have been y units of biomass per year per unit area, or that trawl-caught species richness is generally on the order of z species, etc. From these and similar kinds of observations, we can set bounds on total system removals, aggregate (guild, trophic level, taxonomic groups, etc.) biomass allocations, and related emergent system properties. The role of this type of management advice is to evaluate long-term trends and explore generalized "what-if" scenarios, particularly to elucidate whether a suite of policy objectives is simultaneously feasible, and if not, what tradeoffs (in general terms) may need to take place. This advice can be either qualitative or quantitative, but information is such that the latter is doable and certainly preferred. The point of this advice is to take a broader view of the system and note major changes that could affect multiple facets of the ecosystem. For instance, if there is an El Niño event, the immediate production capacity of a system may be altered. Another example is, again, the two-tier quotas that have been or are currently employed in various places (e.g. Brown *et al.* 1976; May *et al.* 1979; Constable 2001; Witherell *et al.* 2000), whereby the sum of the harvest for individual species must fall below a total systemic cap, with some allowances for endangered species, habitat protection or uncertainty. I think that strategic management advice is generally underutilized in living marine resource management, yet I also think the role of strategic management and related advice is very amenable to a decision-theoretic framework and is in many respects the heart of EBFM.

Tactical management is probably most familiar to those working in marine fisheries management. The typical reference points and control rules in fisheries management form a well-defined and well-established, highly quantified decision-theoretic framework. This type of advice is typically viewed as binding in scope and forms the basis for specific, short-term management actions in an ecosystem, usually targeting a sector of the fishery that emphasizes one species. We need to continue using and providing tactical management advice, but we must extend the underlying information in such advice to be more inclusive (as appropriate) at the stock level and additionally develop analogous advice at the multispecies, aggregate biomass and systemic levels. Several approaches show promise for implementing tactical EBFM advice at these levels (e.g. Collie and Gislason 2001; Constable 2001; Jamieson *et al.* 2001; Link *et al.* 2002a; Collie *et al.* 2003; Sainsbury and Sumaila 2003).

Clarifying which type of advice or information is being provided will be important for advancing EBFM. When it comes to ecosystem issues, too often heuristic information is given when resource managers want tactical advice and when strategic advice is actually needed. In subsequent chapters I note at which level the scientific information can be used or what type of management context is being considered.

As an aside, it is especially important to note the distinction between heurism and the other levels of management advice. Scientists often want to study a subject ad infinitum until they know for certain its underlying mechanisms. Scientists are effectively trained to want to bore into greater and greater detail, ultimately waving away as much uncertainty as they can before posing their hypotheses and answers to those hypotheses. However, in a natural resources management (NRM) context, managers need guiding principles, best available information, the most up-to-date data, and straightforward advice in order to help make a decision (sensu Rosenberg 2007). Often that decision is made on what is scientifically a less than fully known situation or process, but a decision is required nonetheless.

I highlight this tension for three reasons. First, I am an advocate for doing science to the point that we need to have a buffer between the science and management to allow for a bit of freedom for scientists to explore, speculate, innovate, and hypothesize without some urgent crisis du jour breathing down their necks. For much of the supporting elements and information necessary to do EBFM, we need to let the science develop without such constraints. Second, given the prior point, many scientists need to realize that not everything worth studying is necessarily critical, relevant, or germane to NRM of LMRs. More so, not every degree of detail is needed when providing management advice and we as scientists need to develop skills that provide information that is straightforward, succinct, and on-point to the issues facing EBFM. That is, a multipronged hypothesis with a complex "if-then" scenario is not going to help further EBFM. And finally, sometimes in an EBFM context we simply have to make a decision now, based on the best available information. We need to recognize that we can always revisit a decision in an adaptive management style (Walters 1986) as we obtain more information (see Chapters 11 and 12). But waiting until we have perfect knowledge of every ecosystem function and process does a disservice to the scientific discipline, the management application based thereon, and ultimately the LMRs we are managing.

DECISION-THEORETIC FRAMEWORK

There have been several recent prescriptions for ecosystem management in general, including eight recommendations (NRC 1999), eight guidelines (FAO 2003), six policies (NMFS 1999), seven pillars (Lackey 1998), six elements (Sissenwine and Mace 2003), nine mechanisms (Ward *et al.* 2002), eleven doctrines (Ascher 2001), twelve steps (Ward *et al.* 2002), five principles (Ward *et al.* 2002), five different (Dale *et al.* 2000), six (Costanza *et al.* 1998), seven (Mangel *et al.* 1996), eight (NMFS 1999), eleven (FAO 2003), twelve (Malone 2000), or seventeen principles (Garcia *et al.* 2003), and so on. Shucks, we even have ten commandments (Francis *et al.* 2007) or ten myths (Murawski 2007), depending – I suppose – upon how certain one feels about these

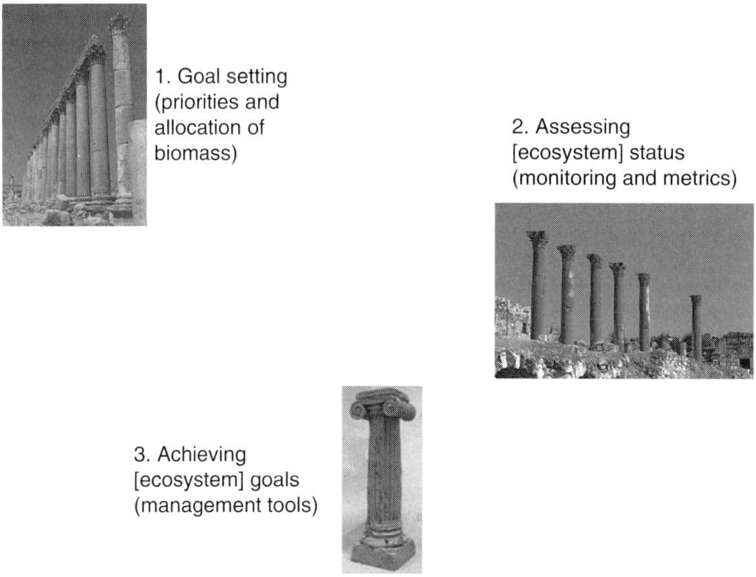

Figure 4.1 A common three-step decision theoretic. These represent "pillars" of any management system.

prescriptions. Embracing the principle of parsimony, I submit that generally most decision-theoretic frameworks consist of three main components (Figure 4.1).

First is to establish goals. What is it that we want from, or out of, an ecosystem in terms of goods and services for society? What are our policy objectives? Central to EBFM will be rectifying competing interests in the allocation of harvestable biomass in an ecosystem. It will be critical for scientists assigned to this mission to note which ecosystem configurations are even feasible (Larkin 1996; Link 2002b).

Second is to assess the system. Where are we relative to where we want to be? This is akin to the way we assess particular stocks. Assessing marine ecosystems requires novel methodologies and means of integration, in addition to a broader interdisciplinary approach. Ultimately, fisheries agencies need to produce an overall ecosystem assessment (i.e. ecosystem status report; NMFS 1999; Link and Brodziak 2002; Boldt 2008; NOAA CBO 2006; Levin *et al.* 2009, EcoAP 2009), updated routinely, to detail and report on the status of an ecosystem. This is analogous to what is typically done for targeted stocks and protected species.

Third is the decision criteria appropriate (reference points, directions or surfaces, thresholds, limits, targets, etc., matched to control rules, action triggers, etc.) to evaluate what steps need to be taken,

Table 4.2. *Combination of the three-step decision-theoretic framework (Figure 4.1) with the gradient of approaches and the three levels of advice used in LRM management*

The capital letters represent various metrics/parameters (or vectors) that would be used to inform the process (in progressive order); the lowercase letters represent similar types of metrics/parameters along the gradient that are less distinct than the capital letters; the numbers represent what would need to be reestimated in each step of the decision framework (proceeding in order, such that A precedes B or C or D, and D1 precedes D2 which precedes D3).

	SS	MS	Agg	Sys
What do we want the system to look like? GOALS				
Heurism	C1a	C1b	B1	A1
Tactical	D1a	D1b	E1	
Strategic			F1	G1
What does the system actually look like? STATUS				
Heurism	C2a	C2b	B2	A2
Tactical	D2a	D2b	E2	
Strategic			F2	G2
What do we need to change to get from where the system is to where we want it to be? MANAGEMENT				
Heurism				
Tactical	D3a	D3b	E3	
Strategic			F3	G3

Note: SS, single stock or species; MS, multispecies; Agg, aggregate biomass groups; Sys, (Eco)System.

if any, to achieve our goals. That is, how do we get to where we want to be? This simple, three-step decision-theoretic framework is conducive to the implementation of EBFM.

Combining the three-step decision-theoretic framework with the gradient of approaches and the levels of advice can be complicated, but upon further inspection it is actually quite instructive (Table 4.2). Doing so also reflects a simplified version of management strategy evaluation or management procedures (e.g. Butterworth and Punt 1999; Smith *et al.* 1999; Sainsbury *et al.* 2000; Butterworth 2007; see also Chapters 10 and 12), which in effect, forms this iterative cycle with a more explicit operating model in the second step noted above. If one realizes that the numbers in Table 4.2 represent the sequential order in the three-step framework, it is clear that measurement and information presented at any level of the gradient or any level of advice needs to address the goals, status, and decision criteria, in that order. Similarly, given the emphasis on a systemic perspective espoused in this book, the information measured and reported would be prioritized at A through G, in order. It may be that indicators or metrics in A–C and even F–G are estimated, monitored/modeled, or reported less frequently than those factors represented in D–E.

The method I am proposing is to set A1, B1, etc., through G1 as appropriate and as data, modeling outputs, etc. are available, but emphasize the systemic end of the gradient first. That is, we would need to have a general understanding (heurism) of the system (A) before we go to MS or SS (C) levels. This understanding would then be a prerequisite for establishing tactical or strategic goals (D–G). After the goals are established, then, on an appropriate timescale, I propose to measure or model A2, B2, etc., through G2. If any values are distinct from the first and second steps, then the approximate difference would result in appropriate management actions to move the system state back to the stated goals (e.g. the difference between G1 and G2 would equal G3, [or D3, E3, etc.]). Since the heuristic level would inform the goal-setting process and the status of the ecosystem, A through C are included in those steps of the framework. However, since this type of advice is not used to invoke management decisions, A through C are not included in the third step.

WHAT TYPE OF ADVICE?

Smith (1998) and Smith and Link (2005) note the difference between policy-forming and policy-supporting advice. For example, much of this book could fall in the category of policy-forming advice; that is, providing scientifically based suggestions of how to develop an improved (management) system to better manage LMRs, typically based on new or rethought scientific theories. Policy-supporting advice would be scientific information that, in the context of an extant management system, provides the basic material from which decisions are made. These distinctions are regardless of whether the advice is heuristic, strategic, or tactical.

I want to be careful not to throw the baby out with the bathwater, nor attempt to prescribe a be-all, end-all policy-forming set of advice for making EBFM globally applicable. The third section of this book explores the implications of policy-forming advice in greater detail, expounding upon general principles that implementers of EBFM may choose to consider. Here I simply want to note that to execute EBFM, much of the extant LMR management systems are set up to incorporate augmented policy-supporting advice. The point is that we rarely augment our advice to include broader ecosystem considerations.

Let me end this chapter by reiterating a previous point. To really do EBFM or EAF (or even more so, EAM or EBM), there is a recognized need to be holistic, coordinated, and integrated. That is, we have had "stovepiped" organizations, whereby the SS models provide policy-supporting advice at a tactical level, which has served as the main emphasis and perspective for most fisheries institutions, without a broader consideration of other factors (often from different disciplines). What I am calling for is to ensure that we have a systemic perspective included in the process as well. To some degree this may very well be viewed as policy-forming advice. Yet without this additional emphasis, we will remain mired in a myopic SS approach, unable to address the core issues of EBFM and, in particular, tradeoffs within a system. We need a SHIRCS (systemic, holistic, integrated, relative, coordinated, and simultaneous) perspective as much as we do a reductionist, SS one. Otherwise we will continue to shirk our responsibility. Or, using Schaefer's (1956) terminology, we need to start emphasizing the third tier as much as we do the first two (Figure 2.3).

So how do we actually do EBFM? We find ourselves in an era of amazing data availability, data collection capabilities, and data storage systems coupled with high-powered computers that are readily available. Thus, much of the debate that occurred over the past century has had several limitations removed. In the next section of the book (II) I provide some tangible, practical, extant technical approaches that we can use to begin doing EBFM.

SUMMARY POINTS

- Taking stock of the key species and processes in an ecosystem, *à la* triage, is a useful first step in evaluating a fisheries ecosystem.

- It is also useful to determine the level of management advice – heuristic, tactical, or strategic – provided in an EBFM situation, and to decide what level is needed.
- There are three basic steps in any natural resources management (NRM) context – setting goals, evaluating status, and management choices – and this simple but powerful framework should be established for the implementation of EBFM.

5

When does it make sense to do EBFM?

I am always doing that which I can not do, in order that I may learn how to do it.

Pablo Picasso (attributed)

INSTANCES WHEN WE WOULD WANT TO CONSIDER EBFM

Are there particular examples where it would make sense to consider a broader, more holistic ecosystem perspective? Here I present some case studies that, combined, may provide some general guidance on the types of situations where using an EBFM would be useful, if not essential. These examples serve as rudimentary cases where EBFM has begun to become operational. In many instances this operationalization was not systematic nor intentional, but a result of being forced to deal with competing interests simultaneously in a given situation.

The example case studies presented here are global in extent – geographically, oceanographically, and scientifically speaking. They cover a broad range of topics, including environmental factors, ecological interactions, multispecies considerations, and full system issues. They are by no means meant to be exhaustive; rather they were chosen to be illustrative of the kinds of instances where EBFM (and more broadly, EBM) is requisite to satisfactorily manage the LMRs in these ecosystems.

As we examine these case studies, which criteria come to mind that are worth noting? How do they address some of the issues raised in Table 4.1? What items might be important (or even critical) to consider for implementing an EAM? Here I will present instances from a SS perspective with ecosystem considerations, and from a systemic perspective with a fisheries emphasis.

ENVIRONMENTAL FACTORS

Particularly in eastern ocean boundary ecosystems (Figure 5.1; Keyl and Wolff 2008), environmental factors can notably influence fish stock dynamics. An example is the Pacific Decadal Oscillation (PDO) and its influence on a range of Pacific salmon species (Beamish 1993; Beamish and Bouillon 1993; Beamish *et al.* 1999) and related species (Hollowed and Wooster 1995; Francis *et al.* 1998; Hollowed *et al.* 2001). In these instances, salmon productivity has been linked to broad-scale climate dynamics. Notable regime shifts have impacted climate, which has altered the productivity of these ecosystems, which in turn has altered dynamics of lower trophic levels. This sequence of events helps to explain nearly a century's worth of cyclic trends in salmon abundance. This example highlights the difficulty of knowing when and how to predict a change in productivity regime, but has also scored one for the Burkenroad interpretation that the environment can be an important determinant of stock dynamics (Skud 1975; Walters and Collie 1988; Rose 2000).

Similarly, the interplay among anchovies, sardines, and the environment is a well-known example (Figure 5.2; Chavez *et al.* 2003). There have been several permutations of this along eastern Pacific ecosystems, as seen in Chile, Peru, the California Current, and parts of the Pacific Northwest (reviewed in Chavez *et al.* 2003; Keyl and Wolff 2008). The general pattern is a "flip-flop" between the spawning stock biomass of the two species, as mediated by both landings and the environmental regime (Chavez *et al.* 2003). This is an interesting case, with what are, in effect, replicates along the coast of both American continents that explore the interplay among environmental drivers, competition between the different stocks, and their interface with overfishing. Several stock assessment models (e.g. Jacobson and McCall 1995; McCall 1996; Jacobson *et al.* 2005; reviewed in Keyl and Wolff 2008) have been established that contain environmental cues to account for potential changes in the oceanic regime and that can influence the dynamics of these stocks.

SPECIES PAIRINGS VIA ECOLOGICAL INTERACTIONS

Oceanographic conditions are not the only ecosystem process suspected of influencing fish stock dynamics. The interactions between seals and cod (and more generally, groundfish) in northeastern Canada have been of interest for some time (e.g. Stenson *et al.* 1997; Bundy 2001; Stenson and Perry 2001; Stenson and Hammill 2006); increasingly so as

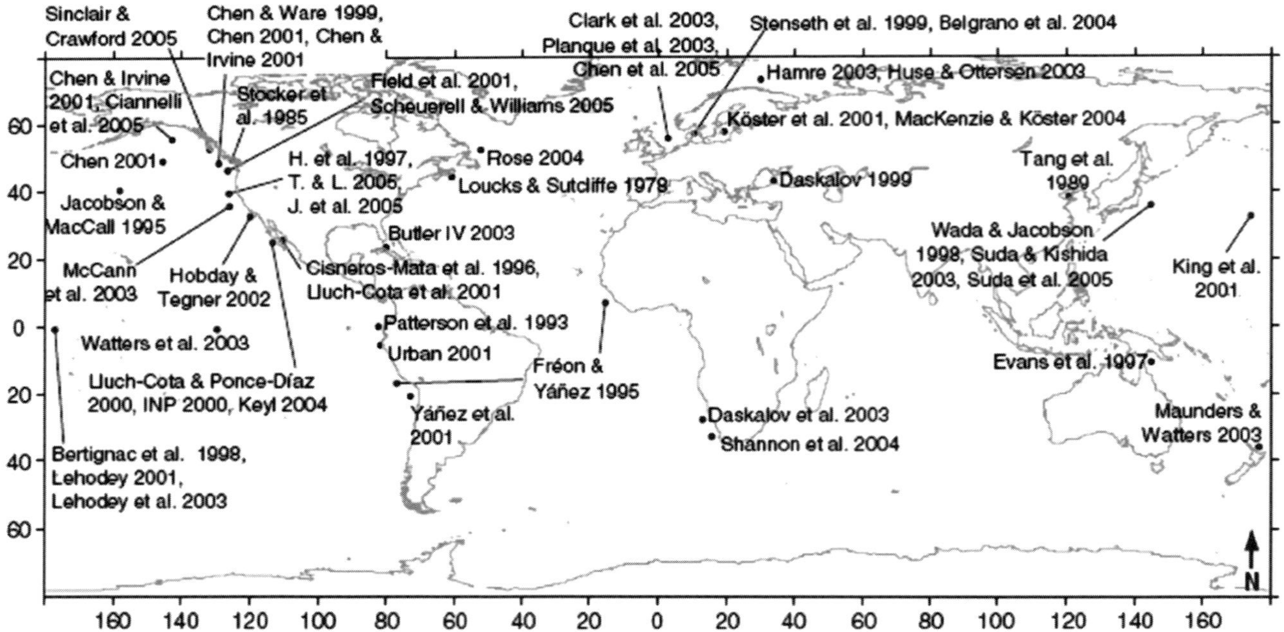

Figure 5.1 Instances where fisheries–climate models have been considered, showing their global distribution. All references noted in the figure can be found in Keyl and Wolff (2008). (Adapted from Keyl and Wolff 2008.)

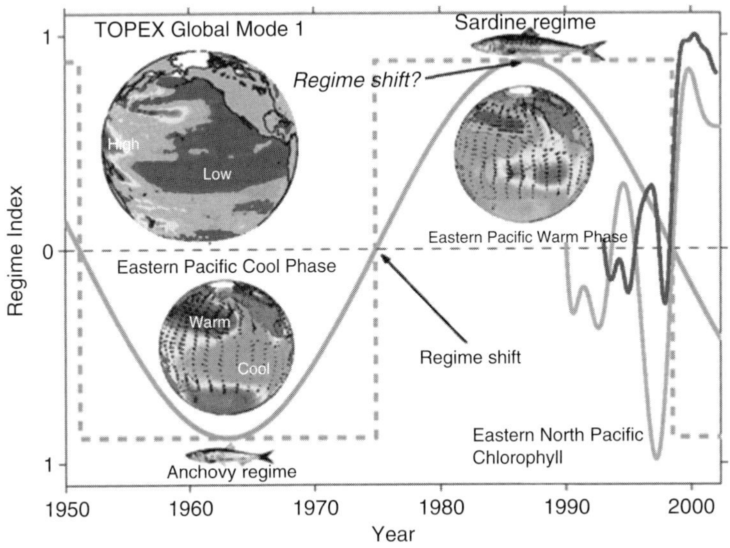

Figure 5.2 The sardine–anchovy "flip-flop" relative to broad-scale climate forcing. (Adapted from Chavez *et al.* 2003.)

explanations for the cod collapse continue to be hypothesized and unsatisfactorily resolved (Smith and Link 2005). The original question posed was whether seal predation (solely or in conjunction with overfishing) caused groundfish stocks to collapse. The evidence is ambiguous but is generally suggestive that this is not the case. The question then became whether seal predation was hindering groundfish stock recovery. Again, the evidence is uncertain, but suggestive that seal predation could be a significant factor impeding the recovery of some groundfish, or at least contributing to their increased natural mortality. This case is an example of exploring tradeoffs among species that have often been addressed under different laws, generally marine mammal species and fish species.

A similar example is the International Whaling Commission (IWC) forage fish issue. There have been several global meetings to address whether or not whales are competing with fisheries that target fishes, namely those which are eaten by the whales (typically forage fishes such as small pelagic fishes; e.g. IWC 2008). As one can imagine, this issue is highly charged both emotionally and politically and has made for interesting media coverage. The general consensus is that despite years of study and modeling, the answer truly depends upon the specifics of any given ecosystem. What has become obvious to even the most casual of observers is that ecosystem considerations are usually invoked to defend either position on each extreme, positions that are established a priori and not necessarily based upon scientific questions. This is an observation that is, unfortunately, all too common when considering EBFM implementation (i.e. a stakeholder says, "Let's do EBFM because it supports what I want. . .").

Another instance of tradeoffs between marine mammals and commercially targeted fish is the situation of stellar sea lions and walleye pollock in Alaskan marine ecosystems (Loughlin and Merrick 1989; Perez 1990; Ferrero and Fritz 1994; Merrick *et al.* 1997; Rosen and Trites 2000). Although walleye pollock is the most valuable and voluminous fishery in the USA (DOC 2008), this stock also has a large number of species interactions, including predation by stellar sea lions. As the sea lions were experiencing population declines, however, by law mitigation efforts were needed. Thus significant regions surrounding rookeries of sea lions were closed off to fishing to minimize both any bycatch mortality and any potential for local competition with the fishery. A suite of models has also been applied to this situation, with the general outcomes broadly ranging but typically precautionary in tenor, with most attempting to confirm or ensure that there is adequate (enough and enough quality) food and safe rookery for these pinnipeds. There have been similar examples in South Africa and elsewhere (e.g. Yodzis 1994, 2000, 2001; Punt and Butterworth 1995; Butterworth and Punt 1999). As in the Canadian seal/groundfish example and in this Alaskan stellar sea lion example, those other models and monitoring can be ambiguous, but it is generally agreed that some coordination and mutual consideration among the different interest groups, stakeholders, and associated managers is clearly beneficial. This is so even if it generally means a more precautionary approach with respect to limiting particular fisheries activities and ensuring positively trending marine mammal trajectories.

Another protected species/targeted species example is the case of North Sea groundfish area closures due to declines in local bird populations (Furness 1999; Furness and Tasker 2000; Tasker *et al.* 2000). These closures were established with a cognizance of the interplay between forage fish, birds, and groundfish, as these forage fish are the nexus through which birds and the groundfish fishery can interact. There was some suspicion that competition between the birds and groundfish (for the forage fish) was impacting the birds within a certain radius of their nesting sites, but also that bycatch in the groundfish fishery was a cause for decline in these bird populations because the birds would get caught while feeding. This case readily shows that the forage fish dynamics can be influenced by environmental factors and, therefore, changes in the distribution of these fishes (as mediated by the

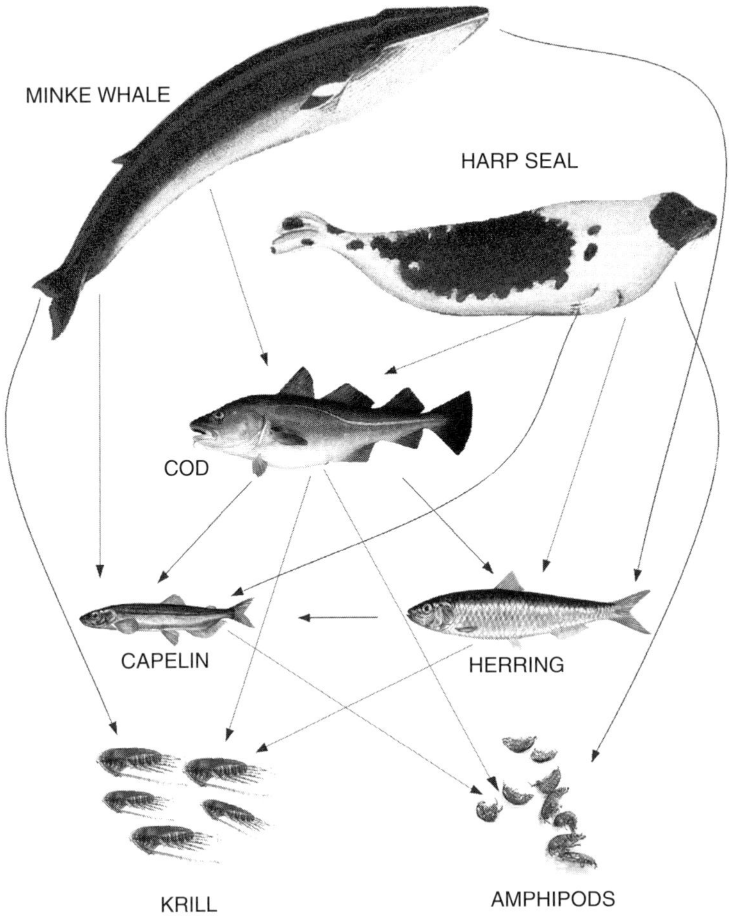

Figure 5.3 Schematic of marine mammal–fish interactions in the Norwegian–Barents Sea. (Adapted from Link *et al.* 2009b; see also Bogstad *et al.* 1997.)

environment) may be reason enough to be adaptive with these area closures and not solely base them on the potential of fishing impacts.

Similarly, the Barents Sea minke whales/cod/capelin example is another situation of tradeoffs among marine mammals, fisheries, and forage fishes, with the twist that there are three species under consideration (Figure 5.3; Bogstad *et al.* 1997; Begley and Howell 2004). The number of possible scenarios to consider increases rapidly, because: (1) minke whales eat capelin; (2) cod eat capelin; (3) minke whales can eat cod; (4) all three species can and often do eat the same micronekton prey (i.e. zooplankton) and thus may potentially compete with one another; and (5) minkes, cod, and capelin all have fisheries prosecuting them. To address this situation, several models have been developed – GADGET, BORMICON, MULTSPEC, etc. (Bogstad *et al.* 1997; Tjelmeland and Bogstad 1998; Stefánsson and Pálsson 1998; Begley and Howell 2004; see Chapter 7) – which explicitly examine and explore the outcomes among the range of tradeoffs resulting from the main species interactions. These models also purposefully contrast environmental effects with fishing, as modulated through species interactions; a useful but rare combination of factors that are considered simultaneously. The outcome of these efforts has been to carefully judge what and how much of each species is harvested and how such harvest could influence the other species of interest.

Instances where broader ecosystem considerations are being considered to explore tradeoffs are not just limited to targeted-protected species situations.

In the Northeast US (NEUS) ecosystem and the Gulf of St. Lawrence in Atlantic Canada, there has been some concern that increased fish populations are eating a large amount of the American lobster (Hanson and Lanteigne 2000; Nelson *et al.* 2003) and related crustaceans (Worm and Myers 2003). In the USA, the concern has been that the recovered striped bass population has exerted an undue predatory pressure. In the Gulf of St. Lawrence, the concern is that cod are similarly removing a large number of lobster. In both instances the mere potential for a predator–prey relationship has translated into the expressed desire to kill more of the predator (amazingly enough, often in conjunction with catch limits concurrently imposed upon the fishery for those predators). Certainly in doing EBFM I would strongly support examination of ecological considerations. Yet the continued pursuit of such considerations also needs to be based on supporting evidence and data. In this particular instance, based mostly upon food habits sampling and a few simple models (Link and Almeida 2000; Hanson and Lanteigne 2000; Nelson *et al.* 2003; Davis *et al.* 2004; Northeast Fisheries Science Center, unpublished data, 2005), predation on lobsters by these species is a low-frequency event and this perception is usually explained by differences (between stakeholders and fisheries scientists) in understanding over the taxonomic resolution of prey (e.g. most feeding has not been on *Homarus americanus* but on such taxa as mantis shrimp, slipper lobsters, other crustaceans, etc., that can have body parts resembling those of lobster).

Another example of tradeoffs with which I am intimately familiar is the dogfish–cod situation in the NEUS (Link *et al.* 2002b). It has been hypothesized that: (1) cod (or really any groundfish of interest) are eaten

Table 5.1. *Example characteristics of stocks especially meritorious of an EBFM*

- highly migratory
- locally dominant in terms of biomass
- are ubiquitous and spatially overlapping with many species
- eaten by protected and commercially valuable species, e.g. odontocetes, seals, birds, fish, invertebrates
- in competition with protected or commercial species, e.g. planktivores, ichthyoplanktivores
- feed on larvae of other commercially targeted species
- have a large fishery potential in their own right
- feed at lower or intermediate trophic levels
- are an important nexus between lower trophic level production and upper trophic level species of commercial, ecological, economic, or legal importance
- have a very high trophic efficiency and relatively strong interaction strengths
- represent a high degree of biomass flux horizontally both within and outside of the system
- have a high linkage density
- have temperature-mediated changes in distribution, migration, or production

by spiny dogfish; (2) cod are outcompeted by dogfish; (3) cod are moved out of an area when dogfish migrate through; and (4) cod, for some reason, just generally don't do well compared with dogfish. Again, I am strongly in favor of considering ecological factors in a fisheries context, but in all instances the ecological evidence (diets, consumption rates, dietary overlaps, population trajectories, recruitment relationships in the presence of dogfish, etc.) does not support these hypotheses. I certainly recognize that the local effects of dogfish schools could influence local fishing conditions for a day or two, but this has not apparently translated into population-level effects. Thus I hypothesize that the differences may be due to differential fishing pressure on these stocks. (I have also hypothesized simply that, for whatever reasons, people don't like dogfish.)

In general terms, there are also particular stocks that have characteristics which make them especially good candidates for a broader, ecosystem approach (Table 5.1). They do not need to have all these characteristics, but often species that possess many of those noted force us to evaluate tradeoffs. Often these are small pelagic species, and as a class of species by themselves should be good candidates for EBFM given all the factors described above.

These examples could really be a pairing of any species that have at least two competing user groups or stakeholders. Often the stark contrast of those pairings coincides with distinct interests and desired outcomes regarding the differential level of abundance of said species. My observations are that the scenarios and outcomes are never as simply binary as people postulate or would like; the classic hare–lynx example (Elton and Nicholson 1942) often invoked is the exception rather than the rule. Thus we need to consider simultaneously a broader set of factors than solely the tradeoffs among simple species pairings.

MULTISPECIES CONSIDERATIONS

One way to consider a broader set of factors, as alluded to in the Barents Sea example, is to explore multispecies considerations and their associated models. For example, there are at least three main predators of Atlantic menhaden: striped bass *Morone saxatilis*, weakfish *Cynoscion regalis*, and bluefish *Pomatomus saltatrix* (Garrison and Link 2004; NEFSC 2006; Garrison *et al.* 2010). A multispecies virtual population analysis (MSVPA) was conducted for menhaden to account for these three main predators. This effort was highly specified (regarding terms of reference for the menhaden stock) and the result is that a key output of the MSVPA, the predation mortality (M2) vector, has been used in single species virtual population analysis (SSVPA) for menhaden.

MSVPAs were originally developed for the North Sea and the Baltic Sea fish communities (Helgason and Gislason 1979; Pope 1979; Gislason and Helgason 1985; Daan and Sissenwine 1991; Magnusson 1995; Vinther 2001). These applications have been varied, but have generally utilized this multispecies approach to explore tradeoffs in biomass, as mediated through predation, among forage fish and groundfish. These cases represent some of the more famous and earliest multispecies modeling efforts in fisheries. Yet they also represent, until more recently, some of the more famous cases where the MS advice (to change mesh size in the first North Sea instance) has been ignored even though the model has been proven to be validated with subsequent sampling and monitoring.

The Convention for the Conservation of Antarctic Marine Living Resources (CCAMLR) serves as the venue to manage LMRs in the Southern Ocean associated with the Antarctic region. In that context, it was one of the first marine LMR organizations to formally adopt an EBFM; it was also one that clearly needed to do so (Constable *et al.* 2000;

Constable 2001). Scientific groups at CCAMLR have established several multispecies models. One for the Antarctic (Constable 2001) contains 13 species or species groups, and includes species interactions, protected species (i.e. marine mammals) and both targeted forage and targeted apex fish species. This effort evaluated incidental fisheries impacts on an ecosystem and specifically sought to evaluate contrasts among different objectives. The resultant reference points (RPs) were mainly based on production models and were generally more conservative RPs when including species interactions. This particular case, although similar to the Barents Sea one, is specifically useful as it provides an example of an aggregate biomass decision criteria framework.

In a recent groundfish assessment review meeting (GARM) for 19 NEUS groundfish stocks, the question of whether simultaneous B_{MSY} was obtainable for all stocks was explored (NEFSC 2008; Gamble and Link 2009). Prior studies (Brown *et al.* 1976; May *et al.* 1979; Beddington 1986) indicated that it would be difficult to have biomass at maximum sustainable yield (B_{MSY}) for all stocks at the same time. From a systemic perspective, the question really became whether there is enough energy and production in the system to support fish stocks all at maximal levels. Several modeling approaches were utilized, with the general consensus that it would be unlikely to have simultaneous B_{MSY} for these stocks. This confirmed prior studies from this ecosystem (Brown *et al.* 1976; Link *et al.* 2008) and related studies elsewhere around the globe (e.g. May *et al.* 1979; Beddington 1986; Mueter and Megrey 2006). This case serves as an example where aggregate reference points were considered and the system production capacity was specifically examined; an example likely needing repetition in many other situations.

SYSTEMIC PERSPECTIVES

The Venice lagoon is situated in the northeast corner of the Adriatic Sea and is, in many respects, the basis of all of the wonderful culture that has been established in Venice over the centuries. Yet the lagoon is facing a host of problems, including the need to keep shellfish and finfish stocks healthy, mitigate organochlorine deposition, maintain an active shellfish fishery, overcome an invasive shellfish, and allow for navigation (Campostrini 2007; Viaroli *et al.* 2007). Yet more noteworthy is the observation that Venice is sinking, and there is a clearly recognized need to keep silt from passing through the lagoon and entering the Adriatic Sea. Related to that is the need to protect the city from storm surges and associated flooding, thus the proposed storm "gates" being constructed at the entrance to the lagoon, a project which also has a potential to impact the lagoon (discussed in Campostrini 2007). This example highlights the need to consider not only tradeoffs among ocean use sectors, but also the cumulative impacts of those many uses. Several models have been or are being developed to explore scenarios that minimize adverse effects across the various tradeoff combinations (discussed in Campostrini 2007; F. Pranovi, personal communication, 2008).

Another example in the northwest Australia Shelf (Sainsbury 1988, 1991, 1998; Sainsbury *et al.* 1997; similar to Figure 1.5) was the first to demonstrate theoretically, as particularly coupled with field data, the importance of area closures to mitigate the effects of fishing. Effectively, closing an area to fishing allowed for the development of increased (biotic) habitat structure, which enhanced the productivity, size, and community composition of commercially desirable fishes. This was also a tropical instance (relatively rare compared with the efforts seen in subtropical and temperate systems in the literature), and the results have not always translated into higher latitude systems (Kaiser and de Groot 2000; Kaiser *et al.* 2002; Kaiser 2003; Link *et al.* 2005a). However, in many respects, it has stimulated a wide range of related research (e.g. Kaiser and de Groot 2000; Pope *et al.* 2000; Kaiser *et al.* 2002; Kaiser 2003; ICES 2003). The work by Sainsbury and colleagues also served as the basis for several subsequent modeling efforts that have led to a multiple use, multiple sector consideration of EBM for that region.

The Great Barrier Reef, particularly as seen through the institution of the Great Barrier Reef Marine Park Authority (GBRMPA), is another Australian example that seeks to explore and address multiuse management challenges (GBRMPA 2008). The key question has been how to keep fisheries and siltation from destroying the reef and associated reef fish. A suite of stakeholders, models, monitoring, and management (especially zoning) has been emplaced, and although the results have varied, it is an example of an institutionalized setting that is deliberately attempting to address EBM across multiple ocean use sectors.

The Gulf of Mexico is another instance where a systemic, multiuse perspective is needed to address key issues affecting fisheries and associated fish stocks therein. In particular, the Gulf of Mexico dead zone (Figure 5.4; Rabalais *et al.* 1991, 2001, 2002b; Turner and Rabalais 1994)

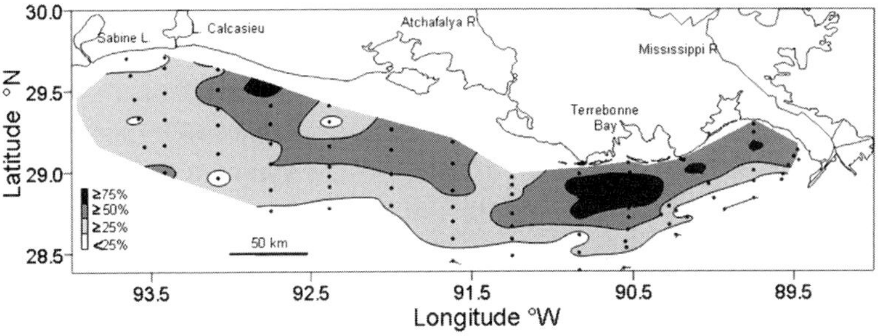

Figure 5.4 Distribution of the frequency of occurrence of mid-summer bottom-water hypoxia over the 60- to 80-station grid (black dots) from 1985 to 2005. This represents the area of low bottom oxygen concentration (dissolved oxygen 2.0) taken from multiple cruises during 1985–2005 in the Gulf of Mexico, specifically giving a sense of the areal extent of the hypoxic zone. (Data and image from N. Rabalais, LUMCOM, as updated from Rabalais *et al.* 2002a, Rabalais and Turner 2006; cf. http://www.ncddc.noaa.gov/website/Hypoxia_Inv/viewer.htm; cf. Rabalais *et al.* 2001, 2002a.)

has negatively influenced fish populations living near the Mississippi River outflow. The Mississippi River runoff creates a hypoxic zone due to its large nutrient load and subsequent biochemical oxygen demand. This hypoxic or anoxic zone kills fish directly, causes them to move to suboptimal habitat, or causes them to move into areas of increased mortality (by both fishing and predation). A key question has emerged as to how best to address upstream and within Gulf of Mexico nutrient loading when it represents a national-scale tradeoff between agriculture and fisheries.

The Laurentian Great Lakes lake trout story also represents a compelling case for examining tradeoffs (Selgeby *et al.* 1995). As a native species competing with introduced salmonids, feeding on an introduced prey base, and being ravaged by an introduced ectoparasite (sea lamprey), lake trout populations were depressed in each of the five Great Lakes. As time progressed, it became apparent that in the five different lakes there were five different responses by this species. A suite of field studies and modeling efforts were executed in each of the lakes, usually emphasizing subsections of species pairings in the context of different fisheries and food webs. These model outcomes generally explained the differential success of lake trout in the lakes due to the vagaries of the food web structures in each (usually coupled with differential fishing pressures). Interestingly, lake trout has been one of the few success stories of stock recovery.

These last few examples highlight the need for taking a systemic view of both fishery production and the factors that can influence fish stocks. It is likely that issues at the far right end of the spectrum in Figures 2.1 and 2.2 are going to continue to surface, and as such a particularly systemic perspective for EBFM is increasingly going to be warranted.

Furthermore, other examples of where EBM has been generally implemented include many US National Forests and Parks, many inland lakes and rivers, the Laurentian Great Lakes, and the Chesapeake Bay (e.g. Yaffee *et al.* 1996; CBFEAP 2006). In many respects, the multiple uses and multiple stressors in these ecosystems have fostered the escalated development of an EBM.

ARE WE THERE YET?

Never rest on what you've accomplished but on what is left to be done.

Jacques Pierre Ribault (attributed)

Recall the questions I posed at the beginning of this chapter. Are there any criteria we can glean from this admittedly superficial presentation of these preliminary EBFM case studies? After a thorough rereading of the literature documenting these cases, a few themes emerge. From all these examples, some recurrent keys to success include: extensive monitoring; clear performance metrics; broad research commitment; addressing pertinent issues at appropriate scales while ignoring non-essentials, but having evaluated a situation to make such a determination; community involvement; adaptive feedback strategies; positive public relations; and sticking to decisions (i.e. following the management procedures process). But more than criteria for successful implementation and consideration, there are the criteria for what these cases have in common such that they would be better served by EBFM as opposed to traditional SS management approaches. The reason these case studies have emerged as requiring an EBFM is summed up in the need to confront the tradeoffs posed in each situation.

Subsequent chapters will address some of the operational aspects associated with some of these (or similar) case studies in more technical detail than I have provided here. But one reason for highlighting these examples now is to emphasize that there is a clear need for EBFM. The other reason is to build upon the themes in prior chapters that doing EBFM is, in fact, feasible. Certainly there are many other examples that could have served as case studies, and if I have left out your favorite example, please accept my apologies. Many of these cases are situations that I have personally worked on or reviewed and are thus situations with which I have some degree of familiarity. Further, I intentionally included some examples where the outcomes are still uncertain, the conclusions are not suggestive of strong ecosystem (particularly ecological) driving forces, or the uptake by managers and stakeholders has not been entirely complete or joyous. I chose such examples to provide a realistic picture of the challenges and opportunities facing us as we continue to explore how best to implement EBFM. A key point from all these examples is that they are ongoing, dynamic situations that at their core require a broader perspective.

In many respects, these examples also represent instances where the scientific and management communities are already taking steps to implement EBFM. Certainly we have not arrived yet and there is much left to do. Yet these examples show that pragmatic application at appropriate places along the gradient of approaches (Figure 2.1) can better address issues that are not capable of being addressed in any other manner besides EBFM.

SUMMARY POINTS

- These case studies have emerged as requiring an EBFM; this requirement stems from the need to confront the tradeoffs posed in each situation.
- These case studies span the range from adding in other considerations to a stock focus, all the way to a full systemic perspective that includes fisheries.
- In some respects, these case studies represent preliminary examples of the beginnings of EBFM.
- There clearly needs to be a more integrated, coordinated, focused effort to fully implement EBFM, and these examples show the feasibility of doing so.

Part II Making EBFM operational: technical considerations

This section builds upon the previous background and context section to provide the quantitative and analytical basis from which one could assess, evaluate, monitor, and provide management advice for an ecosystem. Although I present some of the math behind the thinking of these approaches, my emphasis here has not been to be overly mathematical, but rather to present the material from a conceptual basis. I have tried to present examples of formulae that elucidate and support the concepts, not make them difficult to grasp. As such, the quantitative marine ecologist might find this section somewhat lacking, but I would point such individuals to those references cited for further details. Conversely, the resource managers, stakeholders, or other readers with less appetite for such formulae may find this material a bit more accessible.

Again, my point here (as in the prior section) is to be illustrative, not exhaustive. As such I have provided examples with which I am familiar and that should demonstrate the concepts clearly. By bringing together a wide range of approaches, my intent is to demonstrate that there are methods, approaches, software packages, etc. (i.e. tools) that are extant now which can be used to implement EBFM. The focus of these quantitative tools has been on ecosystem (i.e. biophysical) dynamics, although I do note some models that have socioeconomic elements to them as well. That is, I have focused on LMR and ecosystem dynamics as the primary responses of interest, particularly as they are influenced by a range of driving factors (environmental influences, human activities, etc.). Although the latter driving factors could be modeled and examined as responses in and of themselves, here I mainly treat them as those things which can influence fishery ecosystems.

We start this section by noting the prominence and importance of indicators. How those indicators are used is variable, but they are usually presented in the context of a three-step decision-theoretic framework. I note that they can be empirically based or model based, depending upon the degree of information available. Next is a chapter that explores ESAMs, MRMs, and MS models; I'll let you read on to see what those acronyms mean, but effectively I aim to explore how we can add ecological and environmental factors into existing LMR models and provide improved, ecosystem-based management advice. This advice is usually in the form of modified outputs to the parameters, metrics, performance features, etc., that are commonly used now.

The following chapter then explores a suite of models from an aggregate or systemic view. The thinking, theory, and basic math behind those models is only cursorily presented; the concepts and expected applications are the main focus. How such model outputs are used will be reserved for the next section of this book, but what such model outputs look like and the types of issues they can address are briefly explored in Chapter 8.

The final chapter in this section is on risk assessment and how one can use a wide range of empirical and model-based information in an uncertainty context. This provides a way to integrate a broad range of empirical, highly quantitative, and even qualitative material into outputs that can assist in decision making.

As in the first section, I provide summary highlights for each chapter. Yet here and in the third main section of this book, I also provide some "research remaining" topics to stimulate thinking on these themes. My aim is not to arrogantly direct the research of this disciplinary field per se, but rather to identify those areas I believe need to be "threshed out" to enable a more widespread application of EBFM. Ultimately, this middle section provides the calculation/analytical engine to produce ecosystem-based fisheries management advice, and in the next section we will discuss how that information is or could be used.

6

Ecosystem indicators

The one who says it cannot be done should never interrupt the one doing it.

George Bernard Shaw (attributed)

SYSTEM PERSPECTIVE REQUIRED

As noted in Chapter 4, few ecosystem metrics and model outputs can provide tactical management advice. In truth, it may be unreasonable to expect much tactical ecosystem advice to ever exist. However, there may be some ecosystem analogues to SS reference points that delineate ecosystem overfishing (Murawski 2000; Link 2005; Methratta and Link 2006). Regardless of whether system-level tactical advice exists or not, to readily do EBFM there is a blatant need for an entire system perspective that has been sorely lacking in fisheries science and management. We need to be able to address questions such as whether fishing impacts other ecosystem components and properties (e.g. impacts to other fishing sectors, impacts across broader ocean use sectors, impacts to other, legally considered species), or conversely, whether other ecosystem considerations beyond the fishery are impacting the fishery (i.e. fish stocks). Certainly system-level strategic advice is feasible now.

A useful way to address these questions is to place them in the context of a decision framework and to utilize key ecosystem indicators. As noted in Chapter 4, I submit that most decision-theoretic frameworks consist of three main components (Figure 4.1; Sainsbury *et al.* 2000), the first of which is to establish goals. What goods and services are wanted from or out of an ecosystem for society? What are the policy objectives? And so on. Central to EBFM will be rectifying competing interests for the allocation of harvestable biomass in an ecosystem. It will be critical to scientists tasked with this mission to recognize which ecosystem configurations are feasible.

The second component is to assess the system. Where are we relative to where we want to be? Novel methodologies and means of integration, in addition to a broader interdisciplinary approach, are required to assess marine ecosystems.

The third component is the definition of appropriate decision criteria (reference points, reference directions or reference surfaces, thresholds, limits, targets, etc., matched to control rules, action triggers, etc.) to evaluate what steps need to be taken to achieve the goals. That is, how do we get to where we want to be?

The role of indicators is central to this simple, three-step decision-theoretic framework (Sainsbury and Sumaila 2003) because they permit assessment of the status of a system and because they form the basis, both empirical and theoretical, for the development of reference values. The challenge is to establish ecosystem control rules that prescribe particular management actions if the indicator-based thresholds are exceeded.

INDICATOR TAXONOMIES AND PROPERTIES

There have been several proposed indicator "taxonomies" (Rice 2000, 2003; Rochet and Trenkel 2003; Trenkel and Rochet 2003; Jennings 2005; Rice and Rochet 2005). For example, there is the (Driver) Pressure (Dose, Stressor)-State-Response view (Jennings 2005; Levin *et al.* 2009), the Heuristic-Strategic-Tactical view (highlighted in the last chapter), the Conceptual-Strategic-Operational Objectives (Jamieson *et al.* 2001; O'Boyle *et al.* 2005; O'Boyle and Jamieson 2006), and the System-Response-Performance view (sensu Sainsbury *et al.* 2000; Sainsbury and Sumaila 2003), among others. Despite all the various possibilities, indicators loosely group into *status indicators* and *management indicators*. The key point is that they are used at all of the three steps in an EBFM (or generally in any natural resource) decision framework process (Figure 4.1).

Status indicators assess historical ecosystem trends and influences. This is for targeted, protected, and non-targeted species (TS, PS, NTS, respectively) indicators as well as systemic indicators. The utility of status indicators is in addressing what the recent history was of key process indicators. The objective is to determine the status of the present ecosystem, asking from a policy perspective where we are relative to where we want to be. Providing forecasts for future ecosystem attributes in short to medium terms is useful via these indicators, particularly when exploring the effects of trends or variation in abiotic factors and the effects of biota on one another.

Management indicators help to set thresholds and limits. For instance, what do we know is a precautionary level of action to take given a particular status (FAO 1995; sensu Garcia *et al.* 2003)? Or, even if we are not able to pick the desirable ecosystem state, what states do we want to avoid? More so, these indicators should evaluate performance of past management actions and address the question of where we want to be. By invoking control rules, management indicators suggest what actions need to be taken to achieve objectives. In particular, they help to elucidate what the effects are of alternative policy choices (also called management scenario evaluation, MSE; sensu Sainsbury *et al.* 2000).

So of all the myriad possible indicators available (and being highly suspicious of anyone who proposes the *one* indicator which happens to be his/her area of special expertise), which ones and how many should be selected? At least initially, status indicators should be as inclusive as possible (in order to capture all germane processes in a system). This is a heuristic tool for providing heuristic advice. At this step, these indicators must span the full range of appropriate biological, physiochemical, and socioeconomic factors. Still, the global experience is that a long laundry list of indicators is not immediately helpful for EBFM, particularly when implementing into a strategic or tactical management advice situation.

Several criteria have been established for vetting indicators (e.g. Brodziak and Link 2002; Rochet and Trenkel 2003; Nicholson and Jennings 2004; Rice and Rochet 2005; Jennings 2005). These generally desirable properties of indicators are noted in Table 6.1. Of these, arguably the least important is the issue of precision and measurement error (given the often strategic nature of the issues involved with EBFM). Conversely, the ability to detect change (i.e. sensitivity) and directionality are quite important.

In this vein, there have been exercises to cull indicators from the available range (e.g. Kurtz *et al.* 2001; Link 2001; Link *et al.* 2002a; Trenkel and Rochet 2003; Nicholson and Jennings 2004; Jennings 2005; Rice and Rochet 2005; Methratta and Link 2006). The key observations from these works are that: (1) indicators need to map to major processes and phenomena in ecosystems; (2) indicators need to map to stated (or unstated but legislatively mandated) objectives and criteria;

Table 6.1. *Desirable properties of EBFM indicators*

Directional
Sensitive to change
Range spans natural variability
Precision and variance estimable and reasonable
Unambiguous
Not duplicative nor repetitious
Expressive or representative of key processes

and (3) broad stakeholder involvement in selecting and identifying indicators assists their acceptance later on in the management process (Smith *et al.* 1999, 2008; A. D. M. Smith, CSIRO, personal communication, 2003). Rice and Rochet (2005) in particular have noted some general protocols for the selection of desirable indicators for EBFM. Generally speaking, most examples of indicators selected for EBFM fall into five to seven biological categories, one to three physiochemical ones, and one to three socioeconomical groups (Table 6.2).

Given the types and wide range of information needed to estimate indicators for EBFM, what are the key sources of material for these indicators? In essence, any institution addressing fisheries issues is well suited to calculate EBFM indicators. Both fisheries-independent and fisheries-dependent sampling regimes can provide a proverbial gold mine of data. Additionally, other large-scale monitoring programs not typically utilized in a fisheries context are germane and highly useful here (e.g. satellite imagery, coastal zone management initiatives, water quality monitoring, toxins and contaminants monitoring, etc.). Further, a lot of SS, MS, and ecosystem models can produce these salient indicators. The point is, extant routine and typical fisheries monitoring and modeling can produce much more information if simply reexamined from a new perspective.

PRESENTING INDICATORS

Aside from a long document of individual graphs, there are three main ways one can present an integrated set of indicators in an EBFM context, and they are described below. Whichever is presented depends upon the target audience, level of the issue being addressed, objectives of using the indicators, and similar such considerations (Schiller *et al.* 2001). In all cases, these three presentations

Table 6.2. *A list of the general categories of EBFM indicators usually remaining after an indicator culling exercise*

Biological
Size
Production
Diversity
"Canary" species
Energy flow – trophodynamics
Physical
Habitat
Physiochemical regime
Socioeconomic
Fisheries capacity
Management performance – response

could be used for heurism or strategic management advice; perhaps the latter two could be used for tactical management advice, but generally this has not been done in practice.

First, there is the traffic light approach, whereby percentiles of indicators are presented in a color-coded scheme (Halliday *et al.* 2001; Caddy 2002; Link *et al.* 2002a). No one buys a home appliance or automobile that has consistent and multiple orange or red ratings. Similarly, fisheries managers and stakeholders would like a greener overall system status relative to the history of the system. In redder conditions, caution is heightened, and one could then choose situations to make the overall status of the system greener (accompanied by fuzzy logic models; Halliday *et al.* 2001; P. Fanning, DFO, personal communication, 2002). This approach assumes that mechanisms and specific processes necessary to obtain green conditions are known and manipulatable. Although somewhat qualitative, this approach is feasible for most agencies to use immediately, at least in assessing system-level status.

The second category of presentation includes reference surfaces and polar plots (e.g. Collie and Gislason 2001; Collie *et al.* 2003). Surface plots allow the simultaneous evaluation of three to four dimensions and are often used to show concurrent minima or maxima. Polar plots (compass plots, a.k.a. AMOEBAS) allow the simultaneous evaluation of multiple indicators and are often used to examine biological limit reference points (LRPs) concurrently. Both surfaces and polar plots can be either model or empirically based. The major advantage of this

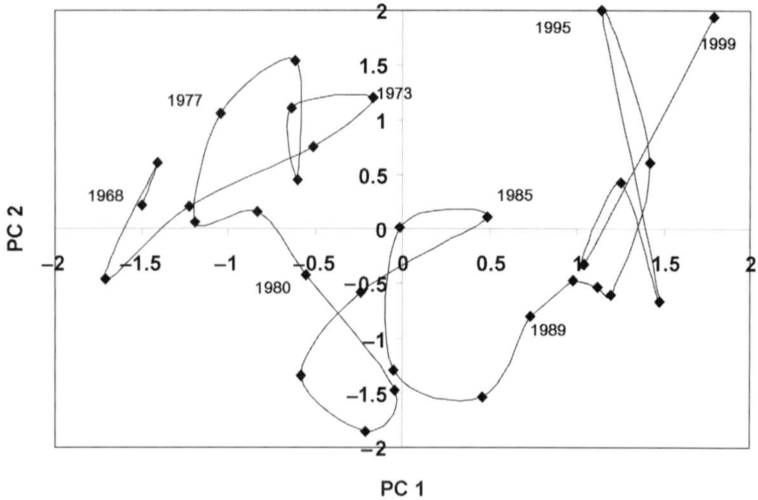

Figure 6.1 Multivariate plot of key ecosystem indicators showing a trajectory of principal component scores through time. (Adapted from Link *et al.* 2002a.)

presentation approach is that it allows for the evaluation of multiple indicators simultaneously. The major disadvantage is that it is often limited to a select set (subset) of indicators, and as such, is not necessarily integrative. Both surface and polar plots can be used to set regions of desirability (e.g. aiming for local maxima, avoiding global minima, bounded within a universal circumference, etc.) in a reference point (surface) sense. Both are particularly useful in evaluating a family of related indicators (e.g. biological LRPs).

The final presentation category is multivariate analyses that present a composite of a larger number of EBFM indicators. For example, the use of common multivariate (MV) tools such as PCA, MDS, etc. can help to reduce dimensionality and help to detect major systemic patterns (Figure 6.1; Link *et al.* 2002a). These methods can provide individual indicator weighting to delineate the major processes acting upon the overall system. These methods are also useful in a culling or vetting exercise. Additionally, canonical analyses (e.g. CanCorr, CCA, RA, DA) can help to elucidate causality between multivariate pressure and response indicators. In one example (Link *et al.* 2002a; Hall and Mainprize 2004), a PCA explains 50% of total variance among the ecosystem metrics. Other studies have suggested that this level of explainable variance may be a common phenomenon among ecosystems (Coll *et al.* 2010, Link *et al.* 2010). But more germane, a MV trajectory confirms and integrates metrics of multiple processes such that a canonical correlation, once established, can elucidate MV relationships and clearly identify the relative importance of different processes. From this, reference quadrats or directions can be established. Although the indicators were empirically based in this case, the question posed an important challenge. Can we get from one quadrat to another by manipulating those causal canonical factors which we can control.

USING INDICATORS

Despite the large efforts in describing and studying indicators, most indicators are not yet usable as reference points in an EBFM context. This is despite clear demonstration that the empirical use of indicators as a function (or partial function) of a stressor (e.g. fishing pressure) can help establish specified thresholds or LRPs (Fulton *et al.* 2005; Link 2005; Methratta and Link 2006). The development of empirically based indicator thresholds certainly needs further work and refinement, but can be used *now* to establish some intermediate decision criteria.

By way of analogy, let us consider the development of MSY reference points and critical toxicity levels. Both of those disciplines (single-species stock assessment [Hilborn and Walters 1992; Quinn and Deriso 1999] or environmental impact assessment [Canter 2001] and ecological risk assessment [Suter 1993]) have developed the requisite machinery and tools to provide appropriate management decision criteria (Figure 6.2a). Both from a fisheries yield perspective for B_{MSY} (50% of carrying capacity relative to fishing) or from a survivability perspective for LD_{50} (50% survival relative to a toxin concentration), the values of interest that are most commonly used represent the results of many years of empirical work, and also correspond to notable inflection points on these curves. The theory and fitting of these curves has followed the empirical observations. I submit that from a systemic perspective, we now could have the analogous empirical basis and will need to seek out similar MV inflection points.

The role of MV reference directions, surfaces, etc. merits further examination and application, not so much in methodology but in terms of how an institution will handle strategic management of a reference direction as opposed to a tactical reference point. For example, it is feasible to use an MV indicator that is a derived reference direction, surface, quadrat, AMOEBA, etc. In this case we need to

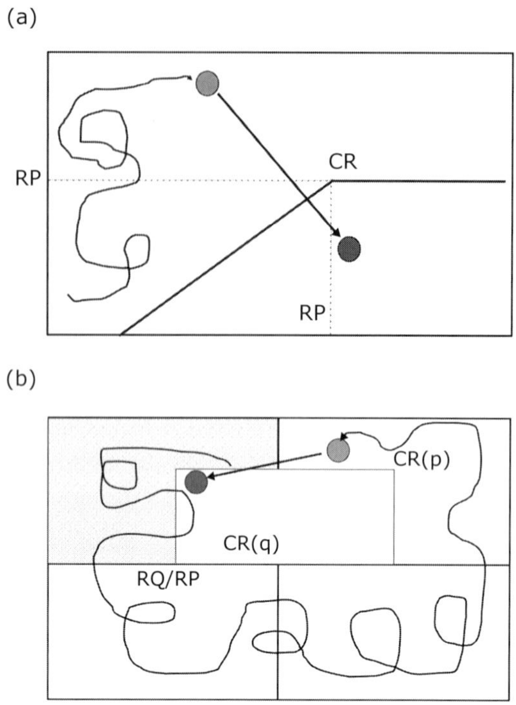

Figure 6.2 Comparison of example single species (a) and ecosystem (b) decision criteria. Both show the current (light gray) and desired (dark gray) status based upon the indicator information, along with a trajectory of the indicator (squiggle line). In the single species instance (a), there are model and empirical-based reference points (RP) and model-based control rules (CR). The action to be taken shows both direction and magnitude. In the ecosystem or multivariate instance (b), there are empirical-based reference points (RP), quadrat (RQ) and directions (arrow), but perhaps relatively more arbitrary or empirical control rules (CR; q, quadrat based; p, point based), despite some modeling to support and further develop these types of efforts (sensu Fulton *et al.* 2003, 2004a). The action to be taken may only show direction unless decomposition of the multivariate axes is undertaken. This perspective does, though, emphasize an integrated, systemic view of the ecosystem.

ask a few questions. What quadrat are we in (e.g. from PCA, or surface, or pole, etc.; Figure 6.2b)? What quadrat do we wish to be in? Irrespective of mechanism, what factors produce the conditions in the desired quadrat (e.g. from CanCorr)? Which of these can we control? Can we then limit effort, landings, etc. for particular aggregate groupings to obtain the desired response? Or do we simply need to wait for a change in environmental conditions? Of course this assumes a reversible trajectory and causality among canonical axes, but it may be a systemic way to evaluate "ecosystem overfishing" (Murawski 2000; Link 2005).

The key concept to be aware of when using EBFM indicators is this: remember the level. What is the issue at hand? And knowing this, what level of scientific advice or information is needed – heuristic, tactical, or strategic? Is the issue really trying to address the effects of fishing on other ecosystem properties or components? Or is it addressing the effects of those on the fishery (namely, fish stocks)? What part of the three-step management process are the indicators being used for? And so on, the point being the need to clarify what the specified needs are in any given situation, rather than attempting to estimate a one-size-fits-all indicator.

REMAINING CHALLENGES

The development of the relationship between indicators and models is growing, but is still in its infancy (Rochet and Trenkel 2003; Fulton *et al.* 2005; Link 2005; Rice and Rochet 2005; cf. Euroceans Indiseas Working Group and SCOR Working Group 119). Further development and translation of indicators into decision criteria will need a predictive element. Underlying models linking "dose response" of selected indicators are required for the broader acceptance of using indicators as decision criteria. Global examples of modeling efforts have matched empirically derived results of indicator thresholds (Rochet and Trenkel 2003; Fulton *et al.* 2005; Jennings 2005; Link 2005), but further exploration is merited. In Chapter 7 I note some EBFM indicators applicable in this context derived from SS and MS models, and in Chapter 8 I note some EBFM indicators derived from aggregate biomass and ecosystem models. Yet again, the ability to move towards delineating ecosystem overfishing requires a SHIRCS perspective.

My main argument throughout this book and particularly in this chapter is that EBFM is feasible *now* with extant information. Assessing the status of an ecosystem is not trivial, but is certainly doable and has been done (e.g. Link and Brodziak 2002; Livingston 2002; Boldt 2008; EcoAP 2009). Again, we need multiple metrics to assess ecosystem status and develop system reference points. We can and have utilized existing MV methods to establish and synthesize relationships and to determine the relative importance among numerous processes in marine ecosystems. We also now know the status of many marine ecosystems (trends, magnitudes, and relationships) in a manner we have never known before.

An example of where this has begun to be done is in the delineation of ecosystem overfishing via a few novel indicators (Tudela *et al.* 2005; Coll *et al.* 2008; Libralato *et al.* 2008). These indicators variously utilize the outputs of common food web or ecosystem models to calculate the probability that an entire ecosystem has been subject to overexploitation. Indicators such as the L-index, probability of sustainability (P_{sust}) or ecosystem-based maximum sustainable catches (EMSC) are calculated using several factors. They include transfer efficiencies; primary production required for the catch; trophic level of catch; and similar information in order to evaluate whether a system as a whole has been overfished, while taking into account energy flows and differential properties of harvesting at different levels of an ecosystem. I suspect this approach will remain quite promising, but will be challenging to implement broadly, as the amount of data and modeling needed may not be readily available around the world.

There are some cautions to be aware of when using indicators. I have argued that both tactical, but particularly strategic, ecosystem reference points (or surfaces, quadrats, directions) exist in an EBFM context. Given this, ecosystem-level management indicators are currently difficult to implement due to stakeholders' and resource managers' lack of familiarity and comfort with them. Indicators alone do not equal reference points, and reference points alone do not equal control rules. Until the hard work of establishing and confirming causality is done, implementing these approaches will have to be in a done in a precautionary manner. Yet the progress we have seen even in the last five to ten years shows that the development and examination of these indicators represents major advancement towards operationalizing EBFM.

SUMMARY POINTS

- Indicators are required at all steps in an EBFM decision framework.
- Multiple indicators are needed to evaluate and manage marine ecosystems and their component biota.
- There is a wide range of ways to present multiple indicators.
- Having ecosystem indicators does not necessarily mean the same thing as having reference points or control rules; translation thereto remains a key challenge.

RESEARCH REMAINING

- Establishing indicators as a function, or partial function, of *F* (fishing rate).
- Translation of indicators into decision criteria (reference points/surfaces/directions and control rules).
- Model predictions of indicator response to a wide range of processes.
- More formalized decision analysis process to evaluate indicators.

7

Expanding the stock focus: what we should have been doing yesterday

If you think you can or think you can't, you're right.

Henry Ford (attributed)

LMR FOUNDATIONS

In the management of any biotic natural resource, it is critical to know how much of that resource exists and is expected to be produced in any given period of time. This is doubly so for LMR.

Hilborn and Mangel (1997) note the difficulties in estimating the abundance of LMR in the ocean, likening it to sampling small mammals in a field or forest from a helicopter by towing a butterfly net. The point here is that our surveying, sampling, and monitoring efforts are foundational in being able to assess the status of LMR, yet they have some inherent uncertainties. Entire tomes have been dedicated to that topic (e.g. Doubleday and Rivard 1981; Murphy and Willis 1996; Anderson 2002; Smith 2002) and I won't repeat the details here. Suffice it to say that at the most basic level, we need a decent set of fisheries-independent and fisheries-dependent sampling in order to do either classical SS fisheries management or EBFM. The call for enhanced global ocean observing systems (IOOS 2007; GOOS 2007) should be one that is fully supported and engaged in by the fisheries community.

Equally critical to LMR management is understanding the key sources that enhance production (habitat, prey field, reproduction, growth, temperature optima, etc.) and cause mortality (fishing, predation, starvation, disease, etc.). That is, what is the rate of growth for a LMR stock (and what factors can alter that) relative to the rate of removals for that stock (and all the factors that can alter it too)? Notably, very few of the extra factors just mentioned are typically considered in most stock assessment approaches. One of the main themes of this book is that they should at least be considered contextually to evaluate the relative importance of *all* processes that can affect LMRs. Said another way, the following should hold generally in all NRM contexts for all renewable resources:

$$\text{Rate of Replenishment} \geq \text{Rate of Removal} \tag{7.1}$$

This simple equation is not earth-shattering, nor novel. Yet we often neglect this basic principle in practice. We often assume a growth rate, and derive management decision criteria based on it to account solely for fishing removals. We often neglect other sources of removals (e.g. predation, unreported bycatch, etc.) and are then mildly puzzled when the growth rates we have based our management on are not sufficient to keep pace with the set of *total* removals.

MORE LMR FOUNDATIONS: MODELS

The most common way to integrate all of the various data and additional considerations noted above is through the use of models. There is a wide range of models available for EBFM (Figure 2.1, Table 7.1), but for now it is instructive to focus on SS models. I noted in Chapter 4 that we should not throw out the baby with the bathwater. Certainly, as implied above, we might want to change the water once in a while (i.e. consider other factors), but in many respects even a rudimentary understanding of basic SS approaches (sensu Hilborn and Walters 1992; Quinn and Deriso 1999) is still warranted as the foundation for implementing EBFM.

Using fisheries as an example, but recognizing that there exist generally similar models and conceptual approaches for other LMRs (PETS, NTS, etc.), there are four main types of stock assessment models. They are: index-based, stock-recruit, production and stage-structured models. I provide a synopsis of each as the foundation for how additional factors can then be incorporated into these modeling frameworks.

Before I delve into these models, however, allow me a quick interlude to describe the rationale and goals for executing these models, often in what is termed a stock-assessment context. The goals of most stock assessments (as seen in standard terms of reference the world over) are typically to:

(1) determine how much of a stock is present in a given ecosystem (be it in terms of biomass, abundance, whatever)

Table 7.1. *Major ecological models used in fisheries*

• Single species
• Single species with add-ons (also known as ESAMs)
• Multispecies (coupled with ESAMS collectively also known as MRMs)
• Food web
• Aggregate biomass
• Habitat
• Biophysical
• Biogeochemical
• Bioeconomic
• Full system

(2) determine how much of a stock is being removed from an ecosystem, mainly via fishing
(3) determine the status of the stock with respect to whether the biomass is at a level above or below what is desirable (based on some biomass reference value, which is usually calculated as part of the exercise)
(4) determine the status of the stock with respect to whether fishing removals are at a level above or below what is sustainable (based on some fishing rate reference value)
(5) recommend further research to address known or perceived information gaps (which are usually ignored and recycled at the next assessment).

The point is, assessment models integrate a wide range of information to produce estimates of biomass/abundance and removal rate, which are then compared to reference levels for each (also estimated via the modeling activity) to determine the status of a LMR. The stock determination is a key outcome of an assessment modeling process; control rules of suggested activity are then enacted to move the stock towards (or maintain a stock at) a desired status. The reference levels in this context are usually known as biological reference points (BRP; see Sissenwine and Shepherd 1987; Restrepo *et al.* 1998; Restrepo 1999; Cadrin 1999; Gabriel and Mace 1999), and are often some form of biomass (B) and fishing rate (F). In the USA, seminal BRPs are typically F_{MSY}, B_{MSY}, or proxies thereof, where those values are the fishing rate and biomass (respectively) that produce maximum sustainable yield (MSY).

There has been copious debate over MSY as a decision criterion (e.g. Larkin 1977; Sissenwine 1978; Ludwig and Walters 1989; Mace 2001; Punt and Smith 2001), but as a starting point (with the freedom to apply other, more appropriate BRPs), it serves as well as any other guideline or benchmark (Restrepo *et al.* 1998; Restrepo 1999; Mace 2001). As one can imagine, estimates of B, F, BRPs, and their relationship to one another (including the choice of particular BRP) can be of significant interest to stakeholders from a particular ecosystem. Although potentially of interest, getting into those debates will be tangential to the point of this chapter. Suffice it to say that how one estimates the values of B, F, and BRPs is, then, quite variable depending upon a host of factors.

The general approaches to develop these estimates fall into one of four main categories. First are index assessments, be they from fisheries-independent or fisheries-dependent data. These are used in data-poor situations to provide the gross magnitude of the dynamics and directionality of a stock, particularly in situations where there is a limited set of information or a new fishery. These models are often, in the snobbishness of some modeling cultures, not considered "real" models per se as they are not full analytical models, but rather are typically represented by a statistical fitting (or more often, smoothing) of some time series of survey or landings data. Some catch-curve approaches are also a type of index assessment. Standard outputs of these types of approaches typically include mean (or median, or similar empirical value for a reference period) biomass or abundance; current or recent trends; current status relative to the mean (or whatever empirical reference value) and perhaps an estimate of mortality (M, where $F = M$ is often set as a rule of thumb). These approaches really do not have any predictive capability as they do not truly have any model structure per se, and they are difficult to add in other factors given this lack of predictive and integrative capability. At least in the USA, with this approach biological reference points are set by proxy at the appropriate empirical levels (as indicated above).

Second are stock-recruit-based or yield-per-recruit-based assessments. These models typically require some understanding or assumption of the relationship between spawners (or more precisely, spawning stock biomass, SSB) and recruits produced from the spawners. Main inputs are vector estimates for the two variables, from which an entire suite of recruitment models can be fit. Important here is an estimate of the timing (in terms of life span) of maturity for a particular stock. These models are classically of the Beverton–Holt (1957) form:

$$R = 1/\left(\alpha + {}^{\beta}\!/_{S}\right) = {}^{S}\!/_{\alpha S + \beta} \tag{7.2}$$

or the Ricker form (1954):

$$R = \alpha S e^{-\beta S} \tag{7.3}$$

where R is recruits, S is spawner biomass, and α and β are fitted parameters. Again, there are other modifications to these two main forms (Hilborn and Walters 1992), but the main point is that these approaches provide an estimate of the number (often translated into biomass) and rate of recruits, and an estimate of spawning biomass (slightly distinct from total population biomass). Relatively simple predictions are feasible, but not beyond a very short time frame. As these approaches have some fitted parameters in the exponent or logistic portion of these equations, one can readily see how additional factors could be incorporated. Generally the outputs of these models inform BRP estimation via proxies to B_{MSY} and F_{MSY}, often with SSB and $F_{0.1}$, $F_{\max}$ or $F_{\mathrm{xx\%SPR}}$, respectively (SPR, spawning per recruit).

Third are bulk biomass production models. This class of models is used when a reasonable amount of biomass (or abundance), growth, and landing (and often including bycatch) data is available as inputs. It is recognized, however, that generally none of that information is age- or size-structured, although some forms can have two biomass pools (in fact, an example would be some specialized forms of yield-per-recruit models). The most prominent example includes the Schaefer (1954, 1957) production model:

$$\frac{dB}{dt} = rB\left(1 - \frac{B}{K}\right) - qEB \tag{7.4}$$

where B is abundance (often understood as biomass); K is the carrying capacity; r is the intrinsic rate of increase (i.e. population growth rate); q is the catchability coefficient; E is the effort expended by a fleet to catch the fish; and the term qEB is the equivalent of fishery harvest. As with all models, production models have sparked their fair share of debate when used in a fisheries context (Larkin 1977; Sissenwine 1978; Ludwig and Walters 1989; Prager 1994; Mace 2001; Punt and Smith 2001; Walters *et al.* 2005). This model has been modified to address a broad range of caveats (e.g. Pella and Tomlinson 1969; Fox 1975; Prager 1994), and many software packages of the model have the ability to provide short-term predictions as well. Standard outputs from these types of approaches are estimates of the fitted parameters of growth rates (r);

carrying capacity (K); often-tuned biomass (B; although this is an input) and a derivation from r to obtain total mortality (Z; where $Z = F + M$ and Z is typically discounted to get F by M assumed to be 0.2). As in the stock-recruit form, this is a model structure with some fitted terms, and one can readily see how additional factors could be incorporated. At least in the USA, biological reference points are set as B_{MSY} and F_{MSY} from production models as $K/2$ and $r/2$, respectively.

Finally, there are stage-structured (age or size) models. This class of models is used when abundance (or biomass) and catch both at age (or size) are known, there is clear age (or size) structure to a stock, there are clear stock-recruit relationships, maturity schedules are able to be determined, and there is an ability to provide initial estimation of natural mortality. Examples include VPA, cohort analysis, SCAA, and similar such models (Hilborn and Walters 1992; Quinn and Deriso 1999). Primary input data includes the typical estimates of biomass and catch, but as partitioned into age classes, plus initial guesstimates of natural mortality between ages. Standard outputs from these types of approaches are back-calculated estimates of stock size (i.e. abundance or biomass) at age, mortality (Z, with a derivation to obtain F) at age, and an assortment of diagnostics related to maturity, survivorship, etc. These models also have the capacity for limited short-term predictions. Although not shown here, the functional forms of these models have the ability to incorporate other factors. At least in the USA, biological reference points are often set using back-calculated information from model outputs to obtain B_{MSY} and F_{MSY}.

REALITY CHECK

> The aim of an argument or discussion should not be victory, but progress.
>
> Joseph Joubert (attributed)

The often unstated desire of the majority of stock assessments (or at least assessment scientists) is either: (1) to move towards a stage-structured assessment model for the benefits of increased precision and predictive ability from those types of models; or (2) to add so many stock-specific terms into any of the generic model forms noted above such that the situation effectively becomes one where there are almost as many parameters to estimate as there are input data points. While either is fine in principle (while admittedly stopping before becoming silly in the latter, overstated case), what I have seen is that this search for precision, with a singular emphasis on Schaeffer's first tier, has in some instances led to an unintentional sacrifice of accuracy when third tier factors are entirely ignored, yet are known to be important. Please do not misread this. The trends, trajectories, *F* rates, and similar assessment model outputs may all still be quite germane, but by ignoring Schaeffer's third tier, often the past realized stock size or the predicted future stock size can be orders of magnitude different than if those other factors are included. Would this have fundamentally changed the primary SS fisheries management advice in any case where a stock was deemed overfished or where overfishing was occurring? Not likely. Rather, by including these other ecosystem considerations the advice may have reached such a determination at an escalated rate.

For example, I have repeatedly seen assessments of forage fish where the total removal of the stock via predation was anywhere from five to 500 times more than what was landed by the fishery. As several of those particular stocks had a relatively truncated size structure, it wasn't as if the predators were taking only the 0+ age group. The implications for these assessments were that there must have necessarily been more fish than we had thought in the past and that there would be fewer fish available in the future when accounting for predation. The resultant outcome would be more conservative reference points for the sustainable utilization of these stocks. Instead, the predation information was effectively ignored since it did not fit neatly into an extant stock assessment modeling and review framework. Had this information been included, the dynamics of many of these stocks would have been much better modeled, with more acceptable model fits and diagnostics. Moreover, including this information would have mitigated a wide range of stakeholder conflicts by having such material directly in the assessment materials, allowing it to serve as a partial basis to examine tradeoffs among many different stocks, types of species, and stakeholder sectors. But, so often, fisheries agencies spend an inordinate amount of resources extracting another significant digit or two of precision while ignoring factors that can have a two to four orders of magnitude level of effect. That is, who cares about the third decimal point on an *F* rate when the biomass is off by a factor of five to ten (or more)?

Fortunately, most fisheries institutions are recognizing a need to change from this purely SS focus, and appreciating the need to incorporate broader ecosystem considerations when assessing the dynamics of a fish or marine mammal population.

There are, however, some remaining SS apologists (e.g. Mace 2004; Hilborn 2004; perspective of Eagle in Leslie *et al.* 2008; S. Gavaris, DFO, personal communication, 2007) that consider anything beyond Schaefer's first or second tiers a complete waste of time and resources. I very much appreciate their perspective and in some select instances actually agree that adding in additional information would be counterproductive. I am quite sympathetic to these SS apologists' concerns in situations where adding in environmental or ecological considerations is not going to change obvious management advice (usually of the form, "Hey, we need to reduce fishing.") and is not going to improve our understanding of the situation. Yet at the same time, we cannot continue to foment a solely reductionist perspective of how marine ecosystems, and their inhabiting fish stocks, function due to the strong potential to miss some extremely key stock dynamics. And as has been discussed in previous chapters, we are now compelled to consider more issues than just the status of a particular stock.

EXPANDING EXTANT ASSESSMENT MODELS: SS ADD-ONS

> When there is a hill to climb, don't think that waiting will make it smaller.
>
> George Bernard Shaw (attributed)

If one starts on the left side of the gradient in Figure 2.1, it is clear that adding in ecosystem considerations to extant stock assessment approaches is a reasonably straightforward way to move towards EBFM. Starting here is palatable for a number of reasons, not least of which is the familiarity of model output and the ability for additional ecosystem considerations to be rapidly assimilated into extant modeling tools. These are readily doable now, and in fact are being done on a limited basis (e.g. Keyl and Wolff 2008). After a triage and heuristic exercise (Table 4.1), the salient processes for any ecosystem and set of species should be incorporated into SS models (Table 7.2) to assess more fully (accurately) the status of a stock.

There are two primary ways to extend SS models, also known as extended stock assessment models (ESAMs). First is by incorporating environmental factors. The second is by incorporating ecological factors.

> Is it me, or is it getting hot in here?
>
> Anonymous

Environmental factors are known to affect LMR stocks in multiple ways. Stocks can be influenced by the environment, typically via physical (e.g. thermal dynamics and heat capacity, thermal preferences, flow fields, frontal boundaries, velocity dynamics, etc.), chemical (i.e. salinity, oxygen, nutrients, or organocholorines), or structural (e.g. bathymetry, habitat availability, substrate size or type, biogenic features, etc.) considerations. These and related environmental factors can affect

Table 7.2. *Possible factors that could be included to incorporate ecosystem considerations into extant SS stock assessment models (understood as either extra initial state variables or factors to tune existing parameters)*

- adding predation
- adding competition
- adding cannibalism
- adding oceanographic impacts
- adding climate impacts
- adding habitat considerations
- adding spatial considerations

stocks through: (1) changes in vital rates (growth, reproduction, consumption, respiration, metabolism, etc.) which all ultimately influence stock production; (2) changes in (biogeographical) distribution, which ultimately influence realized local biomass; (3) changes in migration patterns, which also ultimately influence realized local biomass; or, in a few instances, (4) direct changes in mortality, which ultimately influence both production and realized biomass.

The challenge has been to move beyond simple correlations and provide mechanistic models that have a predictive capability. These need to be models able to predict the fish stock dynamics and the trajectory of the environmental factor of interest, as well as the relationship between them, in a validated manner (Skud 1975; Walters and Collie 1988; Rose 2000; Keyl and Wolff 2008). A recent review by Keyl and Wolff (2008) excellently documents and describes attempts to address this challenge. Effectively, the addition of novel terms in the exponential or logistic terms of Equations 7.2–7.4 (or comparable ones for stage-structured models) could be done with extra α or β as a function of, for example, temperature or a broad-scale climate indicator (e.g. PDO, NAO, etc.). Alternatively, the parameters r or K could be similarly modified in a fitting exercise that included these extra variables. The models would then be rerun, producing appropriately modified BRPs.

The case studies regarding salmon and sardines/anchovies noted in Chapter 5 represent some instances where ESAMs are beginning to include such critical environmental driving forces in assessment models. The catalogs of environmentally extended ESAMs documented in Keyl and Wolff (2008), Townsend *et al.* (2008), and Plagányi (2007) all explicitly highlight the feasibility of doing so, and as such, serve as useful collections of additional case studies where a step towards EBFM has already been clearly taken.

One of the key reasons for considering this is the projected role of climate change (IPCC 2007) as it affects LMRs (e.g. Brander 1994, 1995; Drinkwater 2002, 2005; Hare and Able 2007; Link *et al.* 2009b). There are already documented instances of changes to inherent biota of many marine ecosystems that match what would be expected from climate change, particularly atmospheric and water mass warming (Brander *et al.* 2003; Perry *et al.* 2005; Brander 2007; Mueter and Litzow 2008). As climate change continues, the nuances and challenges of predicting the specifics for a particular ecosystem, as stepped down from global climate models, will remain. It is also likely that these specific changes will not necessarily be linear or unidirectional, and as such, will need to be modeled in their own right. Coupling the results or outputs of these models with ESAMs capable of handling nonlinear, multitrended responses will remain a major – albeit doable – challenge for fisheries science.

On a related note, there are some ESAMs that have incorporated habitat considerations, particularly as it affects r, R, or K in a production or recruitment model context (e.g. Sainsbury 1988, 1991, 1998; Hayes *et al.* 1996; Gislason *et al.* 2000; Fogarty 2005). Although few of these have been actually implemented and used to provide assessment advice directly, they have provided contextual information in the designation of essential fish habitat in the USA. They have also been useful in helping select and provide the theoretical underpinnings of some area closures.

Additionally, there are some SS assessment models that are beginning to incorporate spatial structure (e.g. Orensanz and Jamieson 1998; Hollowed *et al.* 2000a; Holland 2002). Although not technically ESAMs, these spatially explicit models recognize sub-stock structure and have attempted to account for movement between different locales and how this could affect stock dynamics. These models are not per se specifically incorporating ecosystem considerations, but one can readily see how their development would be useful in an EBFM context, particularly as the movement of stock biomass between spatial cells may be mediated via environmental factors and particularly as some aspects of EBFM may be more spatially oriented.

> His bark is worse than his bite.
>
> Idiom

Most of the ecologically extended stock assessment models are actually much more straightforward to implement than most assessment practitioners realize, and certainly as compared with environmentally extended stock assessment models. Particularly if one is focusing on predation, what one does, in effect, is to calculate a set of additional removals (in this case, predatory removals of the LMR stock of interest) and treat it as another fleet in a production or stage-structured model. As reviewed by Tyrrell *et al.* (2010), the basic assumptions for this approach are that input information is extant for calculating diets, consumption, and predator abundances in order to sum that information into a total (predatory) removal. Then that vector of removals can be included into an assessment model to provide an alternative view with expanded removals (predators and fisheries). Novel BRPs can then also be calculated. The challenge is to partition the total MSY and mortality (Z) into that available to or from the fishery and to or from the predators. Noting the mortality decomposition:

$$Z = F + M \tag{7.5}$$

where Z is, again, total mortality, F is fishing mortality, and natural mortality M is:

$$M = \mathrm{M1} + \mathrm{M2} \tag{7.6}$$

where M1 is residual natural mortality and M2 is predation mortality. It is apparent that partitioning mortality, and then an associated biomass level, between fisheries and predators is doable, but will result in a different set of BRPs as compared with when M is set to a constant and not directly modeled (Figure 7.1). Including M2 as part of the evaluations of a stock dynamics seems to be particularly germane to species that are forage for a wide range of predators and that support fisheries in their own right (Figure 7.1).

Examples of applying this type of ESAM include changes to BRPs for the Atlantic herring (Overholtz *et al.* 2008), walleye pollock (Hollowed *et al.* 2000b), Atlantic mackerel (Moustahfid *et al.* 2009), and longfin squid (Moustahfid *et al.* 2009) stocks. In each of these instances, M (technically M2) was found to be time variant, age variant, and thus not constant nor equal to 0.2, as is typically assumed (Hoenig 1983). Additionally, these cases (and related examples; e.g. Link and Idoine 2009; Tyrrell *et al.* 2010) collectively demonstrated that: (1) in back-casting there was more of the stock than previously thought and the total combined yield (predators and fishery) was higher than previously thought; (2) after partitioning between the fishery and predators, the BRPs were more conservative (both F and B proxies for MSY) than if predation had not been included; and (3) in forecasting without accounting for predatory removals, the chance of a stock being in an overfished status was greater than when explicitly accounting for predation.

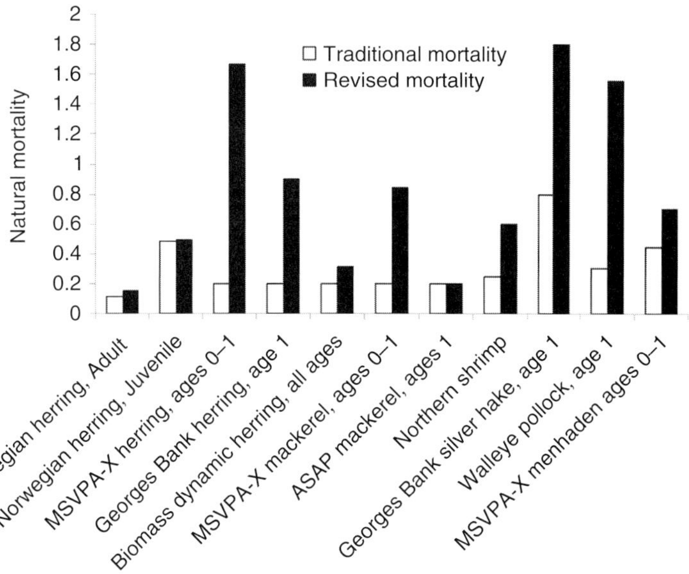

Figure 7.1 Contrast of SS model output biological reference points versus extended stock assessment model or MS model output BRPs. (Adapted from Tyrrell *et al.* 2010.)

Adding predation (M2) as effectively another fleet of removals (consumptive removals) is doable. One could even envision adding cannibalism, similar to how M2 was treated or as an additional term, say, in a stock-recruitment model (Equations 7.2 or 7.3). The point is that expanding the M2 estimates to include marine mammals, cannibals, highly migratory species, birds, or similar species not typically considered in a classical SS assessment context is one way to begin to address, at least in terms of bounding the magnitude of the issue, tradeoffs among different components of the ecosystem. Once the consumptive calculations are done for a particular prey stock of interest, one could reintegrate them from the perspective of a given predator to evaluate the consumptive demand and bioenergetic requirements for a LMR stock, or a stock at a range of levels (e.g. Link and Sosebee 2008).

Adding competition as part of ecologically extended stock assessment models is not as straightforward as adding predation. Although it has been considered (Hollowed *et al.* 2000a; Whipple *et al.* 2000), inclusion of competition has not typically been applied in an operational sense. Mostly this ecological interaction is considered in a multispecies

(MS) modeling context. Yet one could see how it could impact r or K, or be added as a α- or β-modified term in the equations above. The challenge with competition is that there are four requirements that must be fulfilled in order to demonstrate decisively that it is a *possible* consideration: spatiotemporal overlap; similarity of resource utilization (i.e. diet overlap); the existence of limiting resources; and notable population impacts of the interaction (often seen as oppositely trending population trajectories). Because many of the world's marine ecosystems are highly productive, the assumption or condition of limiting resources has tended to minimize the interest of fishery scientists in this particular process. Not to say that competition is not important or does not exist, just that it has not been emphasized nearly as much as predation.

EXPANDING ASSESSMENT MODELS: THE MULTISPECIES WORLD

If, after a triage exercise (Table 4.1), certain ecosystem processes merit consideration, they can be added directly into assessment models. Yet sometimes it is useful to know how a group of populations responds to a process all at the same time. This is the inherent value of multispecies models. These models are not simply a nice research goal; in many ways they are critically needed for better managing and evaluating MS tradeoffs and different user-sectors for fish stocks.

A wide range of reviews and studies have been done on MS models (also known collectively with ESAMs as minimal realistic models, or MRMs) in a fisheries context (e.g. Daan and Sissenwine 1991; Hollowed *et al.* 2000a; Whipple *et al.* 2000; Plagányi 2007; Townsend *et al.* 2008). I won't repeat all of that material here and will refer you to those other works, particularly Plagányi's more recent review on the topic. What is surprising is the relatively limited lack of use and "uptake" of these models (Townsend *et al.* 2008), even though conceptually and mathematically they have been extant for well over 40 years. Some such models have been around for even longer than many of the commonly used SS models (e.g. Lotka 1925; Volterra 1926; Leslie and Gower 1960; Rosenzweig and MacArthur 1963). Simply put, though a large number of community ecologists have explored these types of MS models, they just have not always made the interdisciplinary jump into an applied fisheries context (see Mangel and Levin 2005).

As in the typical SS assessment models, the MS models generally follow the same set of classification as either bulk biomass production or stage-structured models, with a few implementing yield per recruit

(YPR) approaches. Effectively, one can consider them as a set of interlinked assessment models via a submodel that somehow connects (e.g. through species or technical interactions) all the stocks and models them concurrently. Depending upon what factor is to be explicitly modeled (Table 7.2), one of a standard set of MS models can be used (Table 7.3). Below I briefly describe the general MS modeling types and how they can be used in an EBFM context.

Multispecies yield-per-recruit (MSYPR) models (Murawski 1984) are extant but are not all that commonly applied. Basically, they are YPR models for a set of stocks that are targeted together by the same fleet, and linked via measures of "technical interaction" (i.e. bycatch). Key inputs include the standard vectors of *SSB*, *R*, landings, and maturity schedules needed in such SS versions, with the addition of information on the bycatch rates among the species across the fleet. The outputs include the same as those in YPR or some simpler, specialized production models, but with BRPs that are effectively discounted for bycatch.

Multispecies production models are more commonly used than MSYPR models. This group of models is typically useful for addressing predation, competition, or combinations thereof (in conjunction with fishing), without an explicit need for age structure. In some specialized instances (e.g. GADGET – which superseded MULTSPEC and BORMICON; Bogstad *et al.* 1997; Stefánsson and Pálsson 1998; Tjelmeland and Bogstad 1998; Begley and Howell 2004) they are used with targeted and protected species at the same time or can include environmental covariates. The key inputs for these types of models include the standard biomass and landings vectors, but also require some parameterization of the species interaction submodels (e.g. species interaction terms that are fit or initialized based upon data or covariates). Key outputs include the same estimates of *r*, *K*, *Z*, *M*, and associated BRPs (*F*, *B* related to MSY) calculated from them, but are modified to account for species interactions. Recent examples include Bogstad *et al.* (1997), Collie and DeLong (1999), Begley and Howell (2004), Mueter and Megrey (2006), and Gamble and Link (2009), with the number of various applications on the order of ~20–30 globally at any given time. Expanding the classical Lotka–Volterra models in combination with the Schaefer model is a simple way to describe this general class of MS modeling approaches.

Multispecies stage-structured models, particularly multispecies virtual population analysis (MSVPA) and its forecasting companion (MSFOR), are the same as SS stage-structured models, but are joined by a feeding submodel (Andersen and Ursin 1977; Laevastu and Larkin 1981; Daan and Sissenwine 1991). These MSVPAs have been largely

Table 7.3. *Types of models and the ecosystem considerations they address; that is, these categorize major model classes (Plagányi 2007) as they are typically applied to common objectives of model use*

Model classes (Plaganyi 2007)/ Major topics			Extended SS assessment models	Minimal realistic models			Dynamic systems models				Whole Ecosystem models
	Generic model types/Common issues and objectives	Single species	Single species w/add-ons	Multi-species	Aggregate biomass	Food web	Habitat	Bio-physical	Biogeo-chemical	Bio-economic	Full system
Technological interactions	technological interactions		x	x	x	x				x	x
	bycatch	x	x	x	x	x				x	x
Trophic/ ecological interactions	protected species and species of interest	x	x	x	x	x		x		x	x
	commercial fishing on forage species		x	x	x	x				x	x

	tradeoffs among predators being targeted	x	x	x	x					x
	predation of targeted species	x	x	x	x		x			x
Physical/climate drivers	effects of fishing on habitats					x		x		x
	habitat effects on stocks	x	x	x	x	x	x			x
	climate	x	x	x	x	x	x			x
	cumulative effects				x	x	x			x
	toxins/bioaccu–mulation				x			x		x
Spatial features	MPA efficacy, structure, placement					x	x		x	x
	range shifts	x				x	x			x
	habitat restoration strategies					x		x		x
System considerations	invasive species									
	ecosystem health (sustainability, resilience)			x	x					x

Table 7.3. (cont.)

Model classes (Plaganyi 2007)/ Major topics			Extended SS assessment models	Minimal realistic models			Dynamic systems models				Whole Ecosystem models
	Generic model types/Common issues and objectives	Single species	Single species w/add-ons	Multi-species	Aggregate biomass	Food web	Habitat	Bio-physical	Biogeo-chemical	Bio-economic	Full system
	ecosystem status				x	x		x			x
	biodiversity				x	x					x
	underlying system carrying capacity				x	x	x	x			x
	regime shifts							x			

Socio-economic drivers and management	economic issues	x			x	x			x	x
	tradeoffs among fleets			x					x	x
	determining reference points (systemic)				x	x	x	x		x
	determining reference points (SS assessments)	x	x	x						
	cumulative management effects				x	x			x	x

Source: Adapted from Townsend *et al.* (2008).

established in "old school" fisheries institutions and accompanying situations such as the ICES context or in parts of the northern USA where there have been extensive and ongoing data collection efforts. Many of the concerns with the use of MSVPAs have centered around the lack of data to initialize and parameterize these models, and with the assumptions of the feeding models (e.g. type of functional response, how to measure and parameterize suitability, etc.). Modifications to MSVPA have resulted in the 4M and MSVPA-X variants, which address different caveats of the general approach, mostly in the feeding model cognizant of some of these assumptions (Magnusson 1995; Vinther 2001; Garrison and Link 2004). These models are known as data "hogs" in that they incorporate not only all the catch-at-age, abundance-at-age, length-at-age, etc., typical in a SSVPA, but also have extensive parameterization and initialization requirements for the feeding submodels. The key outputs are similar to those in SS stage-structured models, but often include an explicit presentation of M2. In fact, these models were designed to evaluate and estimate age-structured predation impacts, and an output vector or matrix of M2s provides valuable context for ESAMs and classical SS assessment models. The BRPs calculated from MSVPA outputs are, like ecologically extended stock assessment models, generally more conservative to account for predation than their SSVPA counterparts. To my knowledge, there are <10 applications of MSVPAs globally, mainly reflective of the large data requirements of this approach. Recent examples include Livingston and Jurado-Molina (2000), Tsou and Collie (2001a, 2001b), Vinther (2001), Jurado-Molina and Livingston (2002), Jurado-Molina *et al.* (2005), Garrison and Link (2004), Tyrrell *et al.* (2009), and Garrison *et al.* (2010). Incorporating environmental factors into a MSVPA is theoretically possible, but has proven to be not that practical.

As in the ESAM examples, considerations such as spatial structure, habitat effects, or environmental influences can also be incorporated or adopted. Typically, MS models have tended to focus on ecological interactions and omitted those other considerations. But some such as MULTSPEC and GADGET (see Bogstad *et al.* 1997; Begley and Howell 2004), OSMOSE, SEPODYM, and APECOSM (reviewed in Townsend *et al.* 2008) represent a merging of both ecological and environmental considerations in a MS modeling context.

A special instance of MS models is the case of bioeconomic models. Before I continue, allow me to provide a disclaimer for my social scientist colleagues: (1) I recognize that there is a broad range of, and a well-developed suite of, models that address this topic (e.g. Edwards *et al.* 2004; Hennessey and Sutinen 2005); but (2) please cut me a bit of slack here as most biologically oriented fisheries and ecosystem modelers typically ignore economic modeling altogether.

That disclaimer out of the way, these models can be truly MS or in aggregate (see next chapter), but usually have relatively simplified representations of fishing and ecological affects. The main rationale behind these models is usually something to the effect of calculating the optimal economic yield across a wide range of LMR stocks, while accounting for species interactions and species fungibility (i.e. similarity in a marketplace), often ending up with a "blended portfolio" set of recommendations (e.g. Edwards *et al.* 2004). The typical outcome of these models is to conclude that harvest at less than SS MSYs, as supported by other MS modeling outputs, is more desirable, particularly as it usually results in greater economic value in the long run. Again, I apologize for the rather cursory treatment of these models, but I did want to acknowledge their existence and recognize their potential benefit in a MS modeling context. We will revisit socioeconomic considerations in subsequent chapters.

The benefits of these MS models are several. They provide a more accurate representation of the factors that affect LRM stock dynamics; they model key processes simultaneously (as opposed to ESAMs); and the model outputs are still in forms familiar to most fishery stakeholders and managers. The costs of these models are the high parameterization and initialization needs (especially for MSVPA) when including these other factors; debate over the functional form of some ecological relationships (e.g. functional response forms); the reality that they could still miss other factors if focused solely on one ecosystem consideration; and the general concern that managers avoid using these model outputs due to their generally more conservative advice.

USING SS ADD-ONS AND MS MODELS

Most of the models presented in this chapter would be useful for providing tactical LMR management advice at any stage of the decision framework process (Table 4.2; Figure 4.1). The benefit of these models is that they produce outputs that are the same as the common terminology and metrics (i.e. BRPs) currently used by most fisheries management systems. The difference between these and other, more classical SS models is that they represent a more realistic set of processes that can influence the dynamics of a stock and, as such, are closer to reality than if such identified factors are ignored.

In some instances MS models can be used to provide strategic advice, particularly in the third step of the decision framework process (Table 4.2). Often the outputs of a MS model (e.g. M2) are used to set bounds on SS models, if not as inputs directly. Often MS models are used to compliment SS or extended SS models, providing a more formal and rigorous set of boundary conditions than would otherwise be available. Yet again, although MS model output can be directly used in a tactical sense, it rarely is due to concerns over too many sources of uncertainty.

The way the discipline of fisheries needs to proceed for actually doing EBFM is to implement SS add-ons immediately, moving towards MS models as appropriate. The requisite review processes, data collection, validation of models and similar vetting machinery in place for SS models decidedly can, and desperately needs to, be modified to include the models and approaches noted in this chapter.

Finally, it is entirely possible that many of the approaches noted in this chapter can be used as "contextual" assessments to inform the lead assessments until review panels and managers become more comfortable with the behavior of these models. I predict (and fervently hope) there should be an increasing use of these MRMs as lead assessments in the next five to ten years. The point is that there is a continued and growing need for application of these approaches, particularly to explore globally common issues associated with predation on forage stocks that are also fished, and in general to evaluate the impacts of climate change. I also reiterate that there is no one "right" model to use (Table 7.3), and in fact, multimodel inference is preferable if at all possible (sensu Burnham and Anderson 2002). But more importantly, these models need to be used for the appropriate level of management issue (as discussed above) and before their use, need to have adequate data quality (Plagányi 2007; Townsend *et al.* 2008), precisely as I noted at the beginning of this chapter.

SUMMARY POINTS

- Because EBFM needs to consider a whole host of other processes and factors, there are extant ways that we can expand our existing fisheries models to better address the many other issues associated with LMR stock dynamics beyond simply fishing.
- Beyond the classical single species models, there are extant single species models with additions that account for

ecological and environmental processes (e.g. add in M2, habitat-mediated changes to *K*, temperature-toggled *r*, etc.).

- There are multispecies models that specifically address species, technical and, to a lesser degree, environmental interactions which are also extant and increasingly being considered.
- All of these are available, can be used, and need to be used yesterday!

RESEARCH REMAINING

- A broader set of applications is needed, in appropriate instances, for ESAMs that incorporate both ecological and environmental factors.
- A comprehensive review for environmentally extended stock assessment models similar to the one of Keyl and Wolff (2008) would be valuable for ecologically extended stock assessment models.
- A broader set of applications is needed for MS models, in appropriate instances.
- Use of hybrid versions of multispecies (i.e. mostly production with some stage-structured models), given their relatively lower data requirements compared with other MS models, should be notably explored.
- Ecosystem terms of reference need to be inserted into LMR stock assessments, to address both environmental and ecological issues.
- There needs to be recognition (and an accompanying admission by stock assessment review panels the world over) that the models described in this chapter are as valuable and informative as are classical SS models, and often have less uncertainty than is typically assumed.
- An agreed set of review criteria for such ESAMs and MS models is needed (see Plagányi 2007; Townsend *et al.* 2008).

8

A systemic focus: what we can do now

All models are wrong, some models are useful.

George Box (Box and Draper 1987)

BIONIC FISHERIES MODELS

In the last chapter, we saw how extant SS models can be expanded to SS add-on and MS models, maintaining the stock or population emphasis. In this chapter we will explore models that can certainly do some of that, but will primarily emphasize models from an aggregate taxonomy and systemic perspective.

In some instances, after a triage and heuristic exercise has been executed (Table 4.1), it can become clear that a SHIRCS perspective is required to adequately capture system or community dynamics and to adequately address the management issues at hand. As we have noted (Table 7.3), there is no shortage of modeling approaches for the range of issues facing us in the implementation of EBFM. But what are the methods and analytical approaches necessary to facilitate such a broader view and to provide information to support an EBFM decision framework? As we explore the answer to that question, we will also return to some of the themes noted in Chapter 6, elucidating more explicitly some of the outputs from these models as they might be used to delineate ecosystem overfishing.

AGGREGATE MODELS

Imagine an instance where an ecosystem has fish species richness that is so high that one could not begin to even delineate among the species, let alone their demographic patterns and vital rates. Imagine also that these fish are collectively harvested by the same gear or fleet such that this group of fish is effectively captured as an "aggregated" stock. It would make sense to develop management options for them as a group, and in fact doing so might be the only feasible approach.

Or imagine another situation where an ecosystem has a set of clearly defined, distinctly targeted, and otherwise obviously different fish stocks. Imagine also that those stocks have had quite variable dynamics, with major swings in biomass among them over a several-decade period. Now image that in aggregate, even though individual populations have been variable, the total biomass of that entire group (functional guild, taxa group, total of all harvested species, whatever) has been relatively stable over time with much less variability than the individual stocks. Consider also that it would make sense from the perspective of the fishery's market and business plan to develop management options based on that more stable property. If one can imagine those things, then one can grasp the need for aggregate models.

If one recalls the production models discussed in the previous chapter, the essentials are the same, from both a stock or a multiple stock perspective. Typical surplus production models formulations (e.g. Pella-Tomlinson or Graham–Schaefer) can be presented as per Prager (1994):

$$\frac{dB_t}{dt} = (r - F_t)B_t - \frac{r}{K}B_t^2 \tag{8.1}$$

where r is the intrinsic rate of population growth; K is the carrying capacity; B_t is the biomass of a stock at time t; and F is fishing mortality. The same model assumptions, inputs, and outputs hold for this modeling formulation as in the single species application thereof noted in the previous chapter. The only distinction is that instead of modeling biomass for an individual stock, one models the biomass for an entire group in aggregate. The benefit of this approach is that surplus production models are known for their important role in fisheries management, particularly as they have served to help estimate key biological reference points (Larkin 1977; Sissenwine 1978; Ludwig and Walters 1989; Prager 1994; Mace 2001; Punt and Smith 2001). These models are based on relatively simple equations and thus are relatively simple to parameterize (e.g. require only aggregate catch and biomass indices), with model results being relatively easy to present and understand (Hilborn and Walters 1992; Prager 1994). Not to be overlooked, in some circumstances surplus production models may provide more accurate assessments than age-structured models (Ludwig and Walters 1989).

From this modeling approach, one could obtain BRPs such as MSY, B_{MSY} and F_{MSY} for an entire group of targeted LMRs as a group and distinct from the sum of those for each component stock (sensu Figure 7.1). From Brown *et al.* (1976), May *et al.* (1979), Beddington (1986), and several other recent examples (e.g. NEFSC 2008; ICES 2008; Gamble and Link 2009), we know that:

$$\sum \mathrm{MSY}_{\mathrm{SS}} > \mathrm{MSY}_{\mathrm{System}} \tag{8.2}$$

Thus, the need to calculate important parameters and BRPs derived therefrom should not only be apparent, but recognized as quite feasible. Examples of their application include Gamble and Link (2009), NEFSC (2008), and ICES (2008). As noted at the beginning of this section, these approaches are particularly useful in instances of high species diversity (e.g. tropical coral reefs) or where issues of simultaneous MSY are of interest. These models are also beginning to be hybridized with related MS production models, whereby some groups are represented as highly aggregated and others are at a stock-specific or species-specific level (e.g. Bogstad *et al.* 1997; Constable *et al.* 2000; Constable 2001; Collie and Gislason 2001; Mueter and Megrey 2006).

NETWORK MODELS

Great fleas have little fleas upon their backs to bite 'em,
And little fleas have lesser fleas, and so ad infinitum.
And the great fleas themselves, in turn, have greater fleas to go on;
While these again have greater still, and greater still, and so on.

(Augustus de Morgan, *A Budget of Paradoxes*)

Food web and mass balance models, broadly categorized as network models, represent an exceptional instance where community ecology and fisheries science have merged across disciplines. In fact, many of the more recent applications of food web models have been associated with fisheries issues (cataloged in Trites *et al.* 1999; Christensen and Pauly 2004; and Link *et al.* 2005c). Typically, these models are established to evaluate the relative flows of energy (or biomass) to and from LMRs in the context of the broader ecosystem. They allow for the evaluation of species interactions, network structure, descriptions (both theoretical and practical) of the food web within which LMRs function, and testing of various harvest scenarios. Many of the unique "cybernetic" and associated metrics (e.g. Odum 1969; Ulanowicz 1986, 1997, 2004; Ulanowicz and Norden 1990; Ulanowicz and Puccia 1990; Ulanowicz and Kay 1991; Jorgensen 1992) output from these network approaches have advanced ecological theory and could potentially be used in an applied fisheries management context (particularly to define aspects of ecosystem overfishing; sensu Murawski 2000; Link 2005; Coll *et al.* 2008), although they haven't been widely applied for the latter application.

Most applications of this approach in a marine fisheries context can be traced back to Polovina's (1984) development for a Hawaiian Islands coral reef fish community (French Frigate Shoals). Several recent software packages have been developed from this (e.g. EcoNetwrk, EcoPath with Ecosim, Dr. Loop, WAND, and so on; e.g. Christensen and Pauly 1992; Walters *et al.* 1997; Heymans and Baird 2000; Allesina and Bondavalli 2003; Kavanagh *et al.* 2004; Dame and Christian 2006) with different permutations, assumptions, analytical solutions, and treatment of caveats. Qualitative modeling (e.g. loop analysis) has been applied, in addition to the more common quantitative mass balance approaches (Dambacher *et al.* 2002). Inverse modeling solutions have also been attempted to validate whether a mass-balanced solution is universally true or just representing one among many possible local minima (Vézina and Pahlow 2003). Some dynamic simulations based on these networks have also been developed (particularly EcoSim; Walters *et al.* 1997).

An important benefit not to be overlooked when adopting this approach is that it provides enhanced ecological realism in the treatment of a set of LMRs, extended beyond MRMs (Chapter 7). Additionally, such food web models often provide the basis for evaluating and addressing biomass tradeoffs (sensu Larkin 1996; Link 2002b) among the full range of species in an ecosystem. Also invaluable is the observation that these models often serve, through their initialization and parameterization, as a catalog of salient data for a given ecosystem that provides the basis for many subsequent modeling efforts. A challenge of these approaches is that some of the software packages are less than transparent, particularly with respect to balancing and the treatment of detritus (although it is becoming less so). Such lack of transparency has been a concern for those more code oriented in the ecosystem and fisheries modeling scientific community. Another drawback is these models' demand for a large range of data and information for organisms not typically considered in a fisheries context. That, coupled with assumptions over particular forms of key processes (e.g. functional feeding responses) has caused some concern (e.g. Yodzis 1994, 2000, 2001; Punt and Butterworth 1995; Butterworth and Punt 1999; but not any more than is typical of general debates on those processes) over the calibration or validity of these models. Another issue is that the user friendliness of some of these software packages has prompted the spurious use of these models without adequate quality control for data, etc. (but again, that is a generic problem not limited to just this class of models). In my view, none of these concerns are insurmountable, but a key remaining worry is that often the model outputs are atypical relative to classical fisheries model outputs (compared with, say, MS production model outputs). Addressing that concern will be a process involving continued and enhanced presentation; education of stakeholders, scientists and managers as to the context of such model outputs; and time. I estimate 10–15 years for this material to become familiar to the point of common acceptance.

In these ecological network approaches, each node (i.e. a biomass group) requires five main elements. Key inputs for each node include estimates of biomass, production, consumption, respiration, and diet composition. Additionally, for some nodes it is germane to estimate other sources of removals, namely fisheries. Often the challenge has been to initialize or parameterize for what are recognized as some important nodes but which are highly understudied taxa groups (e.g. prawns; krill and similar micronekton; mesopelagic fishes; gelatinous zooplankton; many benthic groups; see Link *et al.* 2008, 2009a). Some of the key outputs include network and cybernetic metrics (as has been noted above; e.g. ecotrophic efficiency of all nodes, but particularly of primary producers, mean trophic level of catch, mean transfer efficiency, and primary production required for catch relative to primary production) as well as partitioning of mortality and the relative (trophic, a.k.a. MTI) impacts of each node. Additionally, these model outputs provide the ability to calculate key biomass and production ratios (Link 2005; Link *et al.* 2008); keystone indices (Libralato *et al.* 2006); and the L-index of systemic overfishing (Libralato *et al.* 2008); all of which could be useful for determining the status of an entire system, or key components thereof, in an EBFM context.

Reference points from these models do not exist yet, although one could readily envision how they could be used. Certainly, estimates of B relative to a B_{MSY}, for example (or multiple MSYs upon rebalancing; Link *et al.* 2009b), could be evaluated from these models. These models can also produce estimates of F and M, but it should be noted that these outputs are proportional, not instantaneous, values, and thus perhaps slightly distinct from the F and M typically considered in SS approaches. Again, many of the outputs from these models (e.g. production ratios, biomass ratios, transfer efficiencies, MTIs, L-index, cumulative production across cumulative biomass by trophic level, etc.) could be established as reference values (sensu Link 2005; Methratta and Link 2006; Chapter 6). But in general, doing so has not yet been widely applied. To be fair, additional applications, calibration, testing, development, sensitivity analyses, exploration of the interlinkage among these metrics (especially relative to fishing), and related model "tweakings" might escalate the uptake of their use as novel reference points, reference surfaces, or reference directions.

BIOPHYSICAL, HABITAT, AND BIOGEOCHEMICAL MODELS

As in the environmentally extended ESAM examples (Chapter 7), we return to issues of space, distribution, habitat, biophysical coupling, and similar topics. Here I will only provide a cursory treatment of these models, as they are in many respects a discipline unto themselves (see review in Plagányi 2007; Townsend *et al.* 2008) and as yet have not been that widely applied in a (marine) fisheries management modeling context.

Biogeochemical models are typically established to evaluate LMR in the context of broader material or elemental fluxes in an ecosystem. Somewhat related to food web models, these approaches seek to place fish in a broader systemic context and are often done to explore the impacts of chemical accumulation modeling (heavy metal or organochlorine toxins). These approaches are not usually executed in a typical LMR context (in terms of setting harvest policy and control rules) but rather in a contaminants-toxicity (or more broadly, ecological risk assessment) framework. These approaches are generally more common in freshwater or estuarine systems (Link *et al.* 2005b) than in large marine ecosystems. Yet I mention them here as they have the potential to impact LMRs, and one could readily imagine that as we move towards EBM with multiple ocean-use sectors under consideration, an understanding of some of those decision criteria – and perhaps cross-use in a more classical LMR context – will be valuable.

Biophysical models, as distinct from SS with environmental add-ons, are those models that have a highly developed set of physiochemical properties modeled, with some thought given as to how that can influence LMRs. General examples of this approach are generic nutrients-phytoplankton-zooplankton (NPZ) models, regional oceanographic models (ROMs in the generic, not specific, sense), water quality models, and so forth (Fennel and Neumann 2004). Typically, these are focused on the physics and lower trophic level processes for the purpose of heurism, but some have begun to be applied towards upper-trophic-level-related issues. Examples of particular model applications that have begun to explore physicochemical influences on LMRs include NEMURO.FISH, OSMOSE, SEAPODYM, APECOSM, CBWQM-FEM and DisMELS (see review in Plagányi 2007; Townsend *et al.* 2008). These models are typically established to explore the effects of oceanography on the distribution and resultant local abundance of LMRs (especially upper trophic levels and large marine mammals) as mediated via thermal preferences, vital rates, sufficient food, or oxygen. Particularly for some of the water quality models in estuaries, these models are often applied in the context of a habitat suitability index (HSI). As such, the outputs are not entirely conducive to use in the classical BRP sense, but one could readily imagine how decision criteria and thresholds could be established from the outputs of these models (sensu Chapter 6), as has, for example, been the case in the HSI context. The typical procedure is to run some of these models and then to couple them, offline, with SS add-ons or MS models to ascertain how the processes they model can influence LMR stock dynamics. Then, the resultant BRPs from those ESAMs or MS models are modeled (e.g. see discussion on climate change incorporation in Chapter 7).

As one might suspect, the data requirement for these types of models exceeds that typically needed for LMR modeling (see Chapter 7). Besides basic abundance and vital rates for the biota, there are many other parameters to be considered. There is a plethora of ROM approaches that alone take many gigabytes to parameterize and tune, plus that many gigabytes of memory to run. The physical parameters, even though perhaps a limited list for these models, can expand exponentially due to the finer spatial resolution found in many of these approaches. In toxicity models, one would need to have a set of inputs not only on the contaminant loadings, but also transfer rates among or across species in addition to mortality levels. And so on. The point is that I recognize that these models are important and will increasingly have a place in EBFM. They simply require an ancillary set of data not typically collected in a NRM context and as such, have not yet become widely applied in the LMR management context.

FULL SYSTEM MODELS

Let us say we wish to model a marine ecosystem. We start with the Sun and end with modeling market dynamics, and attempt to model almost every process imaginable in between. There aren't many people

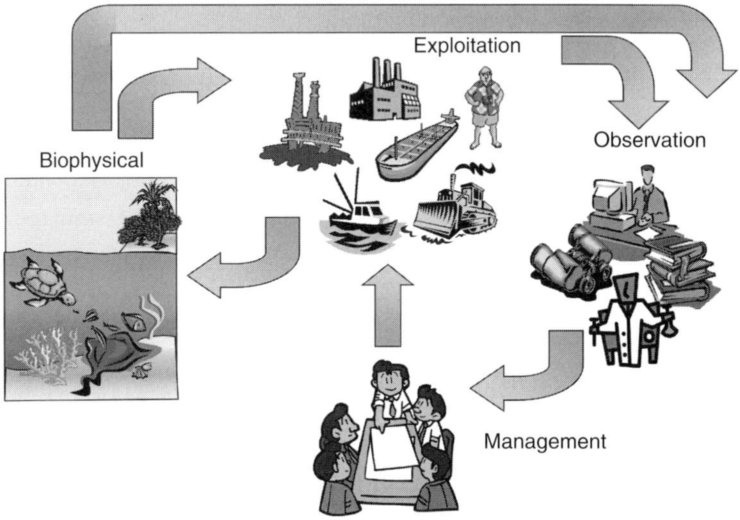

Figure 8.1 Schematic of an example full system model, ATLANTIS. (Adapted from Fulton *et al.* 2007.) Note the ability to simulate the full ecosystem, exploitation, and management process in an iterative manner.

foolish enough to try that, but there are some (I can say that because I am one of them and am friends with many others who are attempting this approach). These full-system models can be thought of as a spatially explicit, dynamic amalgamation of the biophysical, water quality, NPZ, aggregate, bioeconomic, extended stock assessment, MS, and food web models described above and in the prior chapter, coupled with additional socioeconomic models and information (e.g. fleet dynamics, pricing, etc.; Figure 8.1). The purpose of these models is not to confuse (although that is often the accusation leveled against them), but to elucidate the dynamics of an ecosystem while examining *in silico* (that is, virtually) all such dynamics concurrently. The particular utility of these models in an EBFM context is to evaluate the role of LMR relative to the full suite of processes that can affect said LMRs, simultaneously, and also in conjunction with other possible uses of the ecosystem apart from solely the fishing sector.

These models were originally conceptualized a long time ago (e.g. Andersen and Ursin 1977; Laevastu and Larkin 1981), but the computing power and tricks were not at the appropriate level that they are today. Examples of these kinds of models include ATLANTIS and In-Vitro (Fulton *et al.* 2003, 2004a; see review in Plagányi 2007; Townsend *et al.* 2008) and, like biophysical models, their application has been limited in a LMR management context, though it is growing.

By far the most complicated of the models we have discussed (Figure 8.1), the inputs required for these full system models are too numerous to list here; documentation of their structures often results in up to 400-page pieces of gray literature (e.g. Brand *et al.* 2007; Fulton *et al.* 2004a, 2004b). Suffice it to say that most of the data requirements listed in the other models would be helpful to have for these models. The value of this modeling approach, at least as I have seen it in ATLANTIS, is that the functional forms for any given process are recognized as being wide ranging, and thus are not hard coded into the model structure; they are actually set up to have a variable functional form for numerous processes (e.g. recruitment, functional feeding response, sedimentation rate, fisheries selectivity, etc.). This allows one to evaluate and test various model assumptions. The outputs of these models can range from a broad set of ecological indicators to assessment-like BRPs, aggregate group BRPs, network metrics, and so forth. Yet the outputs of these full system models are typically understood to provide heuristic information and strategic advice, as tactical advice is typically obtained via simpler modeling efforts.

USING AGGREGATE AND SYSTEMIC MODELS

Allow me to assert four observations and theses regarding the use of these models. First, they can be used at both a heuristic level and a strategic level (e.g. setting reference directions, bounds of feasible ecosystem states, etc.; Table 4.2). Some versions of aggregate models can be used as composite analogues to their SS counterparts (e.g. B_{MSY} of all flatfish, F for all pelagics, etc.) to provide tactical management advice, but the vast majority of the approaches noted in this chapter should be primarily used for providing a strategic management advice. The point is that these models need to be used appropriately and to provide the right level of management advice.

Second, there is value in multimodel inference (Burnham and Anderson 2002). If the requisite data is available, then it would be wise to construct and apply at least two different modeling approaches to address a particular question (if and as appropriate; Table 7.3) to best elucidate and evaluate the processes under consideration. I reiterate that there is no one "right" model to use (Table 7.3) and having multiple perspectives on the same issue is categorically viewed as a positive. Particularly if multiple modeling approaches concur on some general conclusions as applied to a particular situation, the acceptance of these relatively novel (at least in a fisheries context) modeling approaches and their results will be greater (e.g. NEFSC 2008). Thus, I recommend that wherever possible, this multimodel perspective be adopted.

Third, we should seek to establish at least two tiers of BRPs to account for the limitations of ecological energetics, as noted in Equation 7.2. Primarily by using aggregate modeling approaches, we can and should develop such thresholds (or limits) that account for species and technical interactions, are more stable, and provide the basis for better long-term economic yield (and allow for short-term market dynamics to drive particular species landings). This would be while lower level thresholds (or limits) are applied on a stock basis. This is actually being employed, with some notable success, in the North Pacific Fishery Management Council (Witherell *et al.* 2000; Livingston 2002; Boldt 2008) where a total allowable biological catch (ABC) is set that is less than the sum of SS ABCs. CCAMLR has used a variant of this approach (Constable *et al.* 2000; Constable 2001), and previous ICNAF arrangements were also established to utilize this framework (Brown *et al.* 1976). In all instances this approach provides a built-in level of precaution and allows for some much-needed coordination among the various stocks and fleets that is lacking in a solely SS approach. (Constable *et al.* 2000; Witherell *et al.* 2000; and Sainsbury *et al.* 2000 provide pragmatic examples of instances where these broader factors are being considered in a management context.)

Fourth, we need to establish ecosystem overfishing criteria (sensu Chapter 6) analogous to that of a particular stock. This might mean the development of novel BRPs that are distinct from the straightforward B or F approaches typically used; some examples have been postulated (Coll *et al.* 2008; Libralato *et al.* 2008). Yet without the integrative power of the models described in this chapter, there will be no way to determine whether a system (as a whole) is facing severe depletion to the point which it would require mitigation (or not). Although not directly mandated in the USA yet, this can be implied by the myriad of the country's laws (see subsequent chapters), and is rather explicitly noted in policies found in the laws of other countries and international institutions and treaties.

Generally speaking, the benefits of these aggregate and systemic models are that they provide a more ecologically and environmentally realistic perspective on ecosystem dynamics and their inherent LMR. These models allow one to explore and examine the influences of multiple processes simultaneously, a need that is only going to increase as we move towards EBFM and EBM. Admittedly, the costs of these models are generally higher parameterization and initialization needs coupled with unfamiliarity in a LMR management context; with the associated fears of high uncertainty and often underdetermined solutions. Despite the costs and associated challenges involved in the development, application, and use of these models, there is an ever-increasing set of instances where they are being used around the world. As we move towards EBFM, I encourage their continued adoption.

SUMMARY POINTS

- There are aggregate biomass models which produce analogues to single species model outputs (e.g. $F_{0.1}$ for all flatfish, B_{MSY} for trophic level 4, etc.).
- There is a wide range of network models which produce a wide range of potentially useful metrics.
- There is a wide range of biophysical, habitat, and biogeochemical models which produce a more focused range of potentially useful metrics.
- There are full-system models that produce a lot of information at multiple levels for multiple uses, all of which can provide potentially useful metrics and BRPs.
- Heurism and strategic advice are the primary types of information provided by these system models; only aggregate biomass models and other limited instances may be appropriate for tactical advice.
- There are system-level reference points, directions, and surfaces with associated control rules. The key for successful implementation of EBFM is to expand efforts and use of these reference values.
- Ultimately, fisheries agencies need to produce an overall ecosystem assessment, updated routinely, to detail and report on the status of an ecosystem relative to whether or not there is systemic overfishing.

RESEARCH REMAINING

- A commitment to obtain the diet, consumption rates, and food habits data needed to parameterize these models.
- The same commitment for many of the vital rates of LMR species.
- Research to better establish the links between environmental conditions and vital rates of LMR species.

- A commitment to explore and elucidate those species that are constant information gaps but that are important in many ecosystem modeling efforts (e.g. shrimp, jellyfish, benthos, krill, microbial loop, etc.).
- Comparisons of these applications with their SS stock-based counterparts, as appropriate.
- A broader set of applications, in appropriate instances, for aggregate models.
- Development of better-fitting, diagnostic, and user-friendly tools for many of these modeling efforts.
- A synoptic survey, set of analyses, and review(s) thereof on the sensitivities for many of these models.
- A recognition that in some instances, it is acceptable to let go of a purely stock focus when managing LMRs.
- Establishment of venues to review these models and utilize their outputs, particularly for strategic management advice.
- An agreed-upon set of review criteria for such aggregate and system models (see Plagányi 2007; Townsend *et al.* 2008).

9

Assessing risk: a different view of ecosystem information

Great deeds are usually wrought at great risks.

Herodotus, *The Histories*

HOW DO WE HANDLE LIMITED INFORMATION FOR EBFM?

The previous chapters in this section have all been singularly quantitative, and as such, imply a need for significant quantitative data sets. Yet there are some instances – particularly in many developing countries – where information might be available, but is categorical or qualitative at best. What type of an analytical approach could handle such information? That is, if a situation is truly data poor, can we still implement an EBFM? Could we even start towards EBFM with limited data? Or, what if there is some rigorously quantitative data for some aspects of fisheries science in a particular ecosystem, but limited data on other facets of those factors that can influence fish stocks in that given ecosystem? How would we be able to accommodate that additional information if it were unable to be parameterized into some of the models we have just finished discussing in prior chapters?

I think everyone would agree that moving towards as quantitative an analytical approach as is possible is the most desirable option, and I believe that we should also reiterate *à la* Chapter 6 that most fisheries agencies collect a broad range of data that can, and should, be packaged in unique and novel ways. Also, many of the ocean observing systems now provide online information that is readily available, and we should be as exhaustive and inclusive as possible when establishing indicators and setting up models. Yet in truly data-poor instances, is there an approach that can at least semiquantitatively get us started towards EBFM?

Further, even in instances where we have a data-rich situation and there are copious models, indicators, empirical and observational data sets, and so forth, is there an approach that can allow us to examine all of those factors with some degree of understanding of the associated uncertainty? Is there an approach that can also simplify the outputs of such analytical approaches to facilitate comparisons across stocks, taxonomic groups, geographic regions, ecosystems, or the like, particularly when simultaneously evaluating the degree of probability or certainty of said outputs?

Obviously the answer to these questions is yes, and you may have guessed by the title of this chapter that the answer is risk assessment (RA, or risk analysis, in its more mathematically formal sense). One can legitimately view RA both as a methodology/approach or as a process/framework. Here I focus primarily on the former; I revisit the latter aspects in Chapter 12 (cf. Lane and Stephenson 1998; Peterman 2004; Smith *et al.* 2007).

RISKY BUSINESS

Only by great risks can great results be achieved.

Xerxes (attributed)

We have identified that the need to ascertain the status of fisheries stocks and marine ecosystems is a common issue for LMR management agencies the world over. This need is usually mandated by national or international laws of various forms (see Chapter 11). As we have noted, evaluating the status of stocks has usually been done in a highly quantitative, analytical framework (e.g. Hilborn and Walters 1992; Restrepo *et al.* 1998; Quinn and Deriso 1999) and this is certainly true of evaluating the status of ecosystems (Chapters 6–8). Yet more recently, the evaluation of fish stocks and fisheries ecosystems has increasingly occurred in a risk assessment context (e.g. Francis and Shotton 1997; Lane and Stephenson 1998; Peterman 2004; Fletcher 2005; Astles *et al.* 2006).

There is a wide range of RA methodologies. These include semi-quantitative analyses, because the data necessary for fully quantitative analyses are not available for many fisheries (Dulvy *et al.* 2003). Previous examples of risk assessment have addressed the fishery impacts on bycatch and targeted species (Francis 1992; Frederick and Peterman 1995; Francis and Shotton 1997; Lane and Stephenson 1998; Jennings *et al.* 1999; Milton 2001; Stobutzki *et al.* 2001b, 2002; Peterman 2004), extinction risk (Musick 1999; Roberts and Hawkins 1999; Dulvy and Reynolds 2002; Cheung *et al.* 2005; Mace *et al.* 2008; Patrick and Damon-Randall 2008), and ecosystem state viability (Jennings *et al.* 1999; Fletcher 2005; Fletcher *et al.* 2005; Astles *et al.* 2006; Smith *et al.* 2007). A modified version of the productivity and susceptibility assessment (PSA, a specialized form of semi to fully quantitative RA) has been used in other fisheries (e.g. Milton 2001; Stobutzki *et al.* 2001a, 2001b; Braccini *et al.* 2006; Griffiths *et al.* 2006; Zhou and Griffiths 2008; Patrick *et al.* 2009) and it, or approaches similar to it, has been recommended by several organizations and work groups as a reasonable approach for determining risk to a stock (Hobday *et al.* 2004, 2007; Fletcher 2005; Smith *et al.* 2007). Essentially, these approaches allow for the inclusion of less quantitative information and a wider range of factors, and can complement both stock and ecosystem assessments. The reasoning behind doing RAs is that in data-poor situations they can be very valuable in the absence of a fully analytical stock assessment (sensu Dulvy *et al.* 2003), and in data-rich situations, they can help to prioritize those stocks that should be given more management attention and mitigation (e.g. Peterman 2004; Fletcher 2005; Smith *et al.* 2007).

Risk assessment allows for the evaluation of the risk that a stock or ecosystem will be perturbed beyond what is sustainable (i.e. overfished). It typically uses a categorical (read: semiquantitative) method that has a high utility in its relative ease of use and ability to rapidly identify stocks or ecosystems (or ecosystem components) most at risk. This approach also has the ability to compare across stocks, regions, gears, and fisheries, and, as has been noted, is increasingly being used in a fisheries context.

RECIPES FOR DOING RISK ASSESSMENT

Take calculated risks. That is quite different from being rash.

General George Patton (attributed)

In effect, what risk assessment does is to identify those attributes of the ecosystem that are deemed worth evaluating (see Table 4.1); identify salient features or measures of those attributes that will indicate responses to potential perturbation (sensu Chapter 6); identify those factors that can perturb the ecosystem attributes of interest (see Tables 4.1, 10.1); and then rank or weight or assign probability to the risk of those attributes responding to the driving or perturbing factors such that they end up in a negative or otherwise undesired state (Suter 1993; Canter 2001; Peterman 2004; Fletcher 2005; Smith *et al.* 2007). An example of how one might define levels of risk for various ecosystem attributes is shown in Table 9.1. Interestingly, the degree of risk could also be evaluated for a range of fishing options and strategies for a fixed set of ecosystem attributes in order to determine which "harvest strategies" are less risky to implement.

Again, the particular methodologies for executing RA are quite varied, and range from ranking categorically (e.g. Milton 2001; Stobutzki *et al.* 2001b; Fletcher 2005; Braccini *et al.* 2006; Griffiths *et al.* 2006; Zhou and Griffiths 2008; Patrick *et al.* 2009) to full Bayesian modeling (e.g. Ellison 1996; Punt and Hilborn 1997; Peterman and Anderson 1999; Peterman 2004).

The point is that there are various mathematical methods that take limited information, combine that information, and produce quantitative or semiquantitative (e.g. ranks, categories, etc.) outputs. Ultimately, the outcome of exercising such an approach is a set of outputs that identifies under which conditions, and the relative probabilities thereof, the risk of ending up in an undesirable state might be. Oftentimes these approaches take the form of decision trees that have a probability associated with each step (e.g. Hardwood 2000; Peters and Marmorek 2001; Peterman 2004). The results of RA can also be presented as a plot of risk (vulnerability or susceptibility) versus the ability of an ecosystem attribute to withstand that perturbation (productivity or resilience). An example from the PSA approach is shown in Figure 9.1. Here one could plot any given stock, factor or ecosystem component onto the grid presented to determine the relative risk of unsustainable strategies or stocks most at risk (upper right) versus those that are a lower risk (lower right).

ASSESSING RISK TO AVOID RISK

The main tenet of this chapter is that there is another class of approaches that, although mathematical in their analytical underpinnings, are not solely parameteric nor rigorously quantitative, and are thus able to accommodate a wider range of information. That information is important and would extend beyond what we would classically consider data. I almost did not write this chapter, as much of the process and framework aspects of RA rightly belongs elsewhere in discussions on those topics (Chapter 12), and because much of the

Table 9.1. *An example of summary descriptions of the five sets of consequence levels (i.e. risk) covering the three environmental categories for a generic risk assessment approach in Australia and New Zealand*

The rows (numbers) represent the level of consequence (i.e. risk) and the columns (letters) represent the ecosystem attributes of interest. The content of the cells defines what a consequence to an attribute would look like.

Consequence level	A. Target/vulnerable	B. By-product/other nonretained	C. Nonretained – protected species	D. Ecosystem	E. Habitat
Negligible: no recovery time needed	Undetectable for this population	Area where fishing occurs is negligible compared with where the relevant stock of these species reside (<1%)	Almost none are impacted	Interactions may be occurring but it is unlikely that there would be any change outside of natural variation	Affecting <<1% of area of original habitat area
1. Minor: rapid recovery would occur if stopped – measured in months	Possibly detectable but little impact on population size and none on their dynamics	Take in this fishery is small (<10% of total) compared with total take by all fisheries; species are covered explicitly elsewhere. Take and area of capture by this fishery is small compared	Some are impacted but there is no impact on stock, and this is well below society's acceptable levels	Captured species do not play a keystone role – only minor changes in relative abundance of other constituents	Measurable but localized; affects <1–5% of total habitat area

		with known area of distribution (<20%)			
2. Moderate: recovery probably measured in months to years if activity stopped	Full exploitation rate where long-term recruitment/ dynamics not adversely impacted	Relative area of, or susceptibility to, capture is suspected to be less than 50% and species do not have vulnerable life history traits	Level of interaction/ impact at the maximum acceptable level	Measurable changes to the ecosystem components without there being a major change in function (i.e. no loss of components)	Impacts more widespread but still acceptable; 5–50% of habitat area is affected
3. Severe: recovery measured in years if stopped	Affecting recruitment levels of stocks/or their capacity to increase	No information is available on the relative area or susceptibility to capture; or on the vulnerability of life history traits of this type of species; or the relative levels of susceptibility >50%, and species should be examined explicitly using Set A criteria	Level of impact at above maximum acceptable level. Refer to Set A criteria for any higher levels associated with threatened species	Ecosystem function altered measurably and some functions or components are missing/declining/ increasing well outside historical acceptable range and/or allowed/ facilitated new species to appear	Impact larger than sensible; 20–60% of habitat is affected/ removed

Table 9.1. (cont.)

Consequence level	A. Target/vulnerable	B. By-product/other nonretained	C. Nonretained – protected species	D. Ecosystem	E. Habitat
4. Major: recovery period measured in years to decades if stopped	Likely to cause local extinctions if continues	n/a	n/a	A major change to ecosystem structure and function; different dynamics now occur with different species or groups now the major targets of the fishery	Removal may result in major changes to ecosystem; 60–90% affected
5. Catastrophic: long-term recovery period to acceptable levels will be greater than decades or never, even if stopped	Local extinctions are imminent/ immediate	n/a	n/a	Total collapse of ecosystem processes	Entire habitat in region is in danger of being affected; >90% affected/ removed

Source: Adapted from Fletcher (2005).

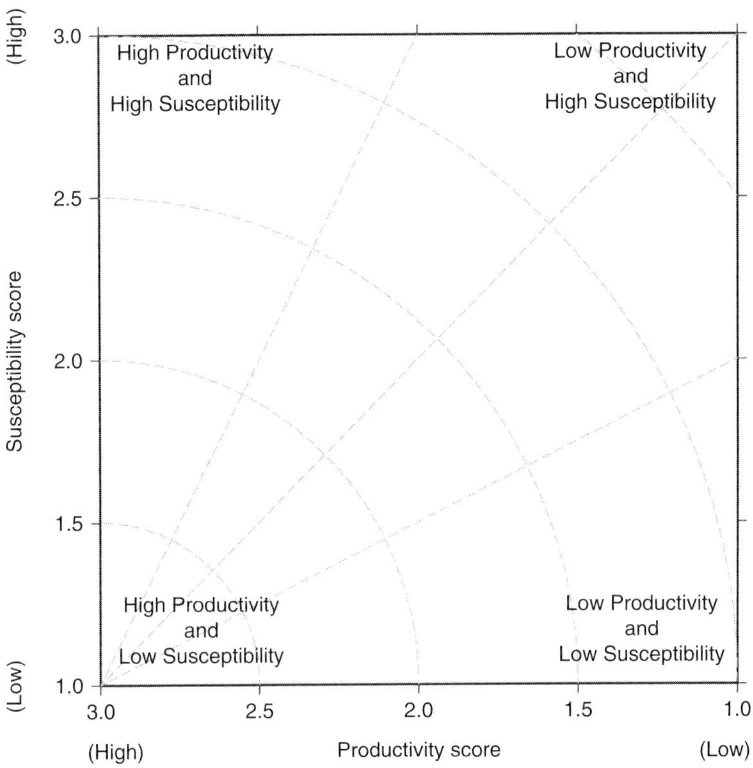

Figure 9.1 An example of the productivity and susceptibility x–y plot. This plot has been adapted from Stobutzki *et al.* (2001b) and modified by reversing the productivity scale to begin with 3 (high productivity) instead of 1 (low productivity).

information and quantitative underpinnings could be (and often has been) subsumed in the more analytical approaches described in prior chapters.

But, the ability to handle a wide range of ecosystem factors, even qualitatively, to the point of ascertaining the relative risk of undesirable states, perturbation, or management strategies, is too important an approach to omit. Risk assessment is an emerging field in this discipline, and certainly has application for EBFM as both a methodology and framework. The chief value of RA is in taking this wide range of disparate information and identifying what will and will not be sustainable ecosystem responses. The outputs from doing a RA are usually rather intuitive and quite informative to NRM managers. As such, they are worth considering as one approach towards implementing EBFM.

SUMMARY POINTS

- Risk assessment (RA) approaches allow for the inclusion of a wide range of information, particularly including that information that, although important, may not necessarily be in the form of rigorously quantitative data.
- Risk assessment approaches are applicable for both data-poor and data-rich situations.
- Risk assessment approaches are fairly flexible, as they have the ability to compare across stocks, taxonomic groups, regions, gears, fisheries, and ecosystems.
- Risk assessment methods are quite variable, and range in their degree of mathematical complexity, but usually result in an output of probable risk to an ecosystem component relative to some perturbation or choice of management strategy.
- Risk assessment is both an approach and a process.

RESEARCH REMAINING

- Case studies and "autopsies" of what worked and what didn't globally, with respect to the range of risk assessment approaches.
- Development (and application) of RA best practices.
- Comparison among the various risk assessment approaches.
- Uptake of risk assessment as part of the management framework and process.

Part III Institutional considerations

The prior section provided an introduction into the technical aspects of how one might assess and evaluate the status of an ecosystem. But then what? Say that one monitors and measures several features of an ecosystem, calculates fisheries independent and dependent indicators, and then models a range of processes and factors. What would one then do with such output? Where would one present those outputs and how would decisions be based on that information? In this section I aim to address questions such as these.

The first chapter in this section explores those facets of the socio-economic context within which fisheries operate. Understanding and characterizing the salient features of any given fishery is a requisite basis for managing that fishery. The next chapter explores governance and management institutions. There I explore the need for such institutions, what roles such institutions should play, and what some desirable features of such institutions might be.

The last chapter in this section attempts to tie the entire book together by discussing a process that explicitly addresses and explores tradeoffs. The emphasis is on a process that recognizes the difference between negotiable and nonnegotiable goals, a process such as MSE that then primarily explores allocation options among the negotiable goals. How the quantitative information is used is discussed there specifically in terms of an operating model, though it is understood to inform the entire process.

Again, my point here (as in the entire book) is to be illustrative, not exhaustive, on the subject. As the topics presented in this section are well beyond my area of expertise, I trust that you will recognize that I present these with a reliance on a mixture of literature, naïveté, and first principles. I aim to present the material for the purpose of furthering the discussion on EBFM. I do so by attempting to compile a broad set of interdisciplinary information into one place, without any pretense of expertise on the topics. As in the last section, I provide both summary highlights and research-remaining suggestions to help the debate along.

10

Why most fisheries biologists become amateur social scientists

It's the economy, stupid.

James Carville [in Bill Clinton's presidential campaign]

The previous chapters described how one might assess ecosystems and their component biota relative to their status, processes, and pressures. The question then becomes, what does one do with that information? How does one utilize it to make decisions? We will address these issues in later chapters. For now we need to take a step back and ascertain just what is it that comprises a fishery and what some of the other factors are influencing the use, harvest, and management of LMRs. That is, how do we view fish stocks and their associated dynamics from a human-use perspective?

To do so, let us ask: What comprises a fishery?

FISHERIES DEFINED

We can define a fishery rather sterilely as the complex of people, their institutions, their harvest and their observations associated with and including a targeted stock or group of stocks (i.e. usually fish), and increasingly, the associated ecosystems that produce said stocks. Let us deconstruct this a bit further. Based on first principles, what are the key components of a fishery?

Fishers are arguably one of the most important parts of a fishery. They are as varied as the fish they catch, and use a variety of gear and vessels to prosecute a fishery. We note that there is a wide range of sectors, fleets, cooperatives, corporations, etc., that all contribute to the dynamics of how and why and when fishers fish. That is, there is a wide range of motives and incentives to fish. Fishers can pursue fishing from a commercial, recreational, or artisanal rationale. I also recognize that fishers are excellent observers of natural history. They are often the front line for observing major processes and events in an ecosystem, which can be both helpful (as a harbinger of major events and as collaborators with scientific observing systems) and challenging (as it leads to misinterpretations from a more singular, or differently scaled, perspective).

Fish processors are also part of a fishery. This is where landings are processed into edible or nearly edible forms (or even nonedible products such as medicines, cat food, etc.). These businesses are where landings are bought from primary fishers and sold to primary (e.g. restaurants, factories) or secondary (individuals) consumers. Fish processors employ many people and establish businesses, which all lead to greater economic impacts for a local economy (ex-vessel revenue, in social science parlance). As such, they usually exhibit a more stable, longer-term economic view than fishers. They are also quite able to adapt to (or in some cases, dictate!) market conditions with respect to the particular stocks being landed, and by implication the particular stocks targetted in an ecosystem.

Fish consumers are also part of a fishery. Very simply these are people who eat fish and they often justify a fishery's existence (at least on larger, commercial or local, artisanal scales). As such, consumers can provide a strong market incentive to modify fisher and processor behavior (e.g. ecolabeling certified fish, taste preferences, etc.) as it relates to what, in what manner, and how much is caught.

Fish conservationists are also part of a fishery. To be fair, most fishers, processors, managers, and scientists have a conservation ethic at heart (see Chapter 3). But this group usually more specifically refers to NGOs and academics who serve as a stakeholder group on behalf of the fish stocks in a fisheries context. Also, to be fair, there is a broad spectrum of NGOs ranging from extremists to centrists who all use differing methods to achieve their conservation goals.

Fisheries managers are also part of a fishery. These are individuals that are somehow selected to collate a broad range of information and then make (informed) decisions about how best to utilize (or optimize?) fish stocks. Fishery managers typically need to address these considerations:

- Economic – i.e. show me the $$$ (or € or £ or ¥).
- Religious. Are there unique beliefs and values that are notably prominent?
- Historical. What has been the precedent and level of expectations?

- Cultural. Where I live is called Cape Cod, not Cape Sculpin. In some towns there are more statues of cod than statues of people.
- Sociological – i.e. basically jobs and related employment opportunities.
- Legal. Numerous laws exist that constrain the range of options (see next chapter).
- Anthropological. How do the six items above interplay in ways that are often unstated?
- Biological. See fish stocks below.
- Political. See money and jobs; the power to influence a situation is an important lure/craving/intoxicant to some people.

Fisheries managers need to balance all of these considerations concurrently as they establish the appropriate harvest policies for a fishery. They are usually answerable to some set of constituents and are often, at least nominally, accountable for the performance of a fishery. They work in a wide range of agencies, councils, and commissions, having entered these positions through election, appointment, or career-conditional hirings (i.e. state, federal, or international civil service).

Fisheries law enforcers are also part of fisheries. Once fisheries managers establish policy, enforcement seeks to ensure that those policies are largely followed. To be frank, enforcement is difficult (particularly in an international fishery). In a domestic fisheries context, port sampling has helped to better measure and address landings, whereas observer sampling has helped to better address catch. Yet fleet coverage by port samplers and observers is expensive, usually <10–20% of any particular fleet. Often fisheries enforcement competes with other law enforcement objectives (e.g. the US Coast Guard has a few other missions to accomplish too), and in practice it is usually only the most egregious and blatant violations that are enforced or addressed.

Fisheries scientists are also part of fishery, particularly as they contribute information to the fishery decision-making process (both policy forming and supporting, sensu Smith and Link 2005). These are "geeks" who study a wide range of issues associated with fish in a (nominally) scientific manner. Their work often involves some combination of monitoring, research, and analysis across a wide range of disciplines, including limnology and oceanography (physical, chemical, and biological); resource economics; social science; ecology (across multiple levels of the biological hierarchy); ecosystems science; systems analysis; statistics; mathematical modeling; operations research; population biology; and stock assessments. I note this range of disciplines, which is by no means exhaustive, to highlight that often, fisheries science is perceived as solely a stock assessment endeavor when there is much more to it. And these disciplines do not include the broad range of support staff and efforts that facilitate the execution of the science. To be fair, even though on paper they like to convey themselves as objective, external observers to a fishery, fisheries scientists can often slide into advocacy for certain positions based upon the information they provide. We will address that topic in further detail below.

Oh yes, *fish stocks* are also part of a fishery. Even ahead of fishers, fish (or more broadly, LMRs in general) are the most important part of a fishery. Without fish stocks, none of the other fishery players noted above would have any role at all in the broader context of society. Think of it this way: without the harvest of LMRs, how many of the activities, efforts, jobs, and economics associated with a fishery would be at the level they are *with* fishery harvest? Sure we would have scientists and NGOs and the Coast Guard, but maybe a lot fewer of them and certainly they'd be working on other things. So in many ways a fishery is a paradox; it is supported by a natural resource that does not have a seat at the table when it comes to discussions about how it is utilized. Without it, however, there would be no institution, and no need for one, to be supported by this resource. Without it, all the associated activities that gainfully employ and otherwise keep occupied a large number of people would also be moot. Note, however, that fish don't vote for you or call you at home or otherwise chew you out if decisions unpopular to their positions or interest are enacted.

WHY WON'T THEY LISTEN TO ME...?

I have provided the preceding rudimentary definition of a fishery for this reason: most fishery scientists are biologically (or more broadly, natural sciences) trained and oriented, yet the majority of issues dealing with a fishery are quintessentially human. If one looks at the list of issues that fisheries managers have to consider (above), the fish stocks are only one (usually minor) factor in their deliberations (e.g. Berkes and Folke 1998; Endter-Wada *et al.* 1998; Charles 2001; Colding and Folke 2001; Franz 2001; Hanna 2001; Holling 2001; Gunderson and Holling 2002; Browman and Stergiou 2005; Pascoe 2006; Rosenberg 2007; Symes and Phillipson 2009), and why biological information is often overshadowed (Rosenberg *et al.* 1993; Beverton 1998; Caddy 1999; Caddy and Cochrane

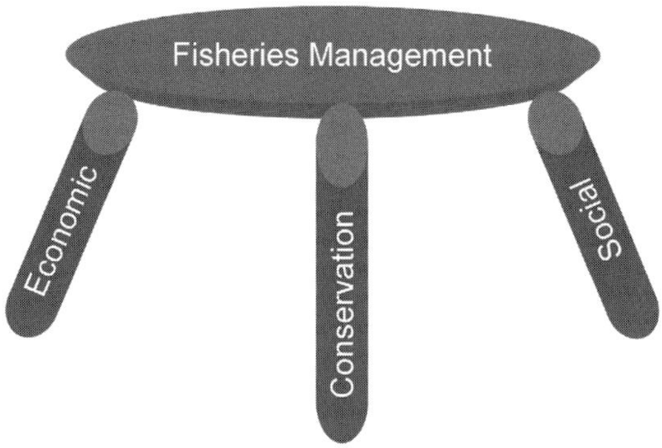

Figure 10.1 The fisheries management "stool" with three component legs supporting the management process, or influencing how it "sits", as the case may be.

2001; Sissenwine and Mace 2003; Caddy and Agnew 2004). All the fishery components noted above are human related, save one. Additionally, in the three-step decision theory framework (Chapter 4, Figure 4.1), certainly the second part is largely biological, but the first and last steps need to explicitly account for socioeconomic factors. In exploring the three legs of the fisheries management "stool" (Figure 10.1), biological factors only come into play in one of the legs, and even then these biological factors are specifically filtered by legal interpretation. Certainly there are legally mandated national and international guidelines or standards if LMR stocks are overfished or endangered from a biological perspective, but commonly those considerations are trumped in spirit, if not practice, by social and economic considerations when hard decisions need to be made. For example, if a fish stock needs to be rebuilt on a 10- or 20-year time frame, everyone can agree to that. But then in implementation, although it would usually make more sense biologically to start as soon as possible, economically or socially (i.e. politically) it might make more sense (at least in the short-term) to plan on the rebuilding at the back-end of that time period to save jobs, ensure current income or value, and see if the status of the resource might somehow change in out-years. This example, and numerous others like it, highlights how clear-cut biological advice can be (and often is) ignored because of other pressures in a management arena.

Because the majority of fisheries scientists have a biologically oriented background, they can be a bit naïve regarding other factors when it comes to the prominence of economic or social considerations in a decision-making process, particularly as those considerations override considerations deriving solely from the status of a LMR stock. I have seen this naïveté initially lead to surprise or shock, then to cynicism or sarcasm, and then (if there is still sufficient enthusiasm so as to avoid fatalism) to an amateur interest in social science. This interest is rooted in the desire to better understand (we are, after all, scientists), explain, and navigate (and perhaps even positively influence) how the NRM process plays out. At the end of the day, fisheries are ultimately all about people – managing people (and their behavior), interacting with people, understanding people, building trust with people, setting goals with and for people – and the fish or LMR or ecosystems that people use are in many ways ancillary to EBFM. It is the wizened, grizzled, ol' veterans in fisheries science and management (and more generally, NRM) that have this figured out (sensu Thomas 2000).

I trust that these observations are themselves not too naïve, are not too late for some in the cynical phase of their career (and an encouragement to them), and sufficiently reach some of the junior fisheries scientists to make them aware of these realities. It would be wise for the discipline of EBFM to explicitly include some aspects of social science as we train the next generation of practitioners. And allow me to humbly admit that it is very clearly as an amateur social scientist (without any claim to wisdom, grizzle, or in any way having the following topics figured out) that I present the remainder of this chapter.

VALUING VALUE

As in the triage effort noted in Table 4.1, a similar exercise is warranted to ascertain the socioeconomic "landscape", or more accurately "seascape", surrounding a particular fishery. The questions posed in Table 10.1 should help to address such considerations. Again, as in Table 4.1, it is not necessary to answer all of the questions in Table 10.1. And it is possible that we may not be able to answer all of them for a given ecosystem in the relatively near future. Rather, we should at least pose these questions as we continue along the EBFM process. By asking the questions, we verbalize and begin to emphasize a novel, broader, systemic perspective. I assert that in so doing, considering broader ecosystem issues becomes viewed as highly amenable, and in fact often aligns with preferences voiced by many stakeholders regarding the socioeconomics issues associated with a LMR NRM context.

Table 10.1. *A general list of issues that need to be considered for evaluating the socioeconomic factors influencing fisheries in an EBFM context*

Economic considerations for EBFM

Ownership

- Who owns the vessels in the fleet?
- Who owns the resource in the ocean?
- Who owns access to the resource?
- Who owns the ecosystem supporting the targeted resource?

Value

- What is the value of vessels, gear, equipment, etc. in the fleet?
- What is the value of the resource in the ocean, as catch, and as landings?
- What is the value of access to the resource?
- What is the value of the ecosystem (goods and services) supporting the targeted resource?
- What are the market forces and dynamics that influence the value of a resource?
- What is the value of the ecosystem (goods and services) supporting the targeted resource?
- How do you calculate or estimate such valuation schemes?

Cost

- What is the cost of vessels, gear, equipment, etc. in the fleet?
- What is the cost of fuel to catch the resources?
- What is the cost to the buyers, processors, and suppliers to get the catch to the market?
- What is the cost to the resource in the ocean?
- What is the cost to the ecosystem (goods and services) supporting the targeted resource?
- How much foregone yield is there due to overfishing, bycatch, etc.?

Fleet dynamics

- What are the operational costs (partially covered above) to the fleet?
- What are the regulatory and enforcement costs to the fleet?
- What are the logistical factors facing the fleet on a broad timescale/average year (e.g. distance to fish, fishing ground hazards/hangs/snags, weather delays, etc.)?
- What are the ecosystem considerations influencing where a fleet fishes?

Optimality

- When, where, and what is the optimal amount of resource, and resource species mix, to catch that balances: (1) a minimization of cost; (2) a minimization of deleterious ecosystem effects; and (3) a maximization of profit?

Risk

- When, where, and what is the appropriate amount of resource, and resource species mix, to catch that balances: (1) a maximization of limiting deleterious ecosystem effects; (2) a minimization of losing profit; and (3) doing so in such a manner that accounts for risk?

Table 10.1. (cont.)

Incentives

- What motivates fishers to fish?
- What motivates fishers to fish in different locales, using different gears, targeting different species, at different rates, etc.
- What motivates fishers to behave in certain ways that are risk prone, averse or neutral?
- What motivates fisheries managers to behave in certain ways that are risk prone, averse or neutral?
- What incentives are there for fishers (and fisheries managers) to be precautionary?
- What incentives are there for fishers (and fisheries managers) to be conservative with respect to catch?
- What incentives are there for fishers (and fisheries managers) to account for ecosystem considerations?

Subsidies

- How much of a fishery is ultimately subsidized (beyond market considerations) by government assistance?
- How much of fisheries science is subsidized beyond a fraction of the total value of a fishery and/or its supporting ecosystem?
- How much does it cost to execute the management process?

Social considerations for EBFM

What kind of a fishery is it

- Commercial?
- Recreational?
- Artisinal?

Culture

- How much of the local history is based on fishing?
- How much of a local/regional economy is supported by fishing (as a percentage of total)?
- How much of a community's or large group of individuals' identity is based on fishing or a fishery?
- How (much) does the local geography, institutions (and names thereof), art, cultural expressions, etc. depict or reflect fishing?

Employment

- How many jobs in a locality/region are based on fishing, both directly as: (1) fishers and buyers; and (2) indirectly as processors, middlemen, consultants, managers?
- What other employment options are available?
- Is entering the fishery sector a viable employment option?

Table 10.1. (cont.)

Power
- What are the main factors that influence where a lot of money and jobs go in a fishery?
- Where does the power in a fishery reside? That is, with whom or what institution, e.g. fishers, processors, managers, conservationists, consumers, enforcement?
- Which individuals are key players in the entire fishery?

Worldview
- What is the social ethos with respect to resource harvesting?
- What is the social ethos with respect to ecosystem considerations?
- What is the mix of stakeholders in a region?

Once a socioeconomic evaluation has been done (i.e. stepping through appropriate aspects of Table 10.1), then it is worth collating (or mapping) a few of these more basic (mainly economic) factors, particularly as it pertains to how most fisheries are prosecuted so as to place ecosystem assessments and related EBFM efforts in an appropriate socioeconomic context. I only provide a cursory examination of them here so as to raise these as important issues, not to provide a definitive treatment for any of them.

There is an obvious need to identify and understand the underpinnings of a fishery, as the social, religious, historical, and cultural drivers are keys to why, how, and when people fish. Knowing the motives of fishers allows one to respect their core values and beliefs while negotiating over management policies. That is, understanding why people choose to fish is critical to understanding why they behave the way they do and why they react in particular ways to certain choices, and allows one to better handle the management (i.e. decision-making) process. An oversimplified observation is that in general, across effectively all fishing communities, fishing is viewed: (1) as more than a livelihood; and (2) usually as a way of life (e.g. sensu Berkes and Folke 1998; Charles 2001; Symes and Phillipson 2009). Managing something that is so central, not only to people's economic prospects, but even to their identity, is something that must be done with, frankly, a bit more care than we typically do.

Furthermore, there are other stakeholders and interest groups regarding LMRs in marine ecosystems. As with fishers, understanding why other people care about fish or whales or their associated oceans is also critical to understanding why they behave the way they do, why they react in particular ways to certain choices, and will allow one to better manage the decision-making process. Clearly not everyone will agree all the time, and there are often legitimate differences of opinion (again, the theme of tradeoffs throughout this book). Knowing that there will be disagreement, though, does not explain why everyone doesn't always agree. But examining a fishery at this level may lead to that understanding, and possessing that understanding will help to resolve a lot of potential conflicts among all the stakeholders. As we move across sectors from EBFM to EBM, the intersections between fisheries and other ocean uses will continue to escalate regarding conflicting valuations and uses of multiple ecosystem goods and services. Where and how we resolve within- and across-sector conflicts is an important question which we will address in the next chapter. In fact, accounting for these types of considerations is, in large part, why the need arises to address different tradeoffs (regarding objectives; Figure 4.1).

The point here is to recognize that socioeconomically driven conflicts (or the potential for conflicts) exist, and understanding the root causes of why they exist will be invaluable for their resolutions. A related point is that an accurate mapping of all the stakeholders interested in a particular ecosystem and its component LMRs is valuable in any NRM context, be it ecosystem based or not.

There is also an obvious and important need to valuate a fishery and its associated ecosystem. Although doing so has been proven to be feasible (Costanza *et al.* 1997; Boyd and Banzhaf 2007; Norton and Noonan 2007), it is often not typically applied to things that have what is termed "existence value", things that are not directly translatable into ecosystem "goods and services". Doing so poses a challenge, especially for issues that most marine ecologists and fisheries scientists had not even been thinking about until recently. For instance, what is a jellyfish worth in an ecosystem in terms of its functional role, sequestration of food that could end up in a targeted fish, service to the ecosystem in transferring lower trophic level energy to protected species like turtles, etc.? Admittedly I am inexpert on the topic, but aside from the inherent difficulties posed by the complexities just noted, sometimes the application of these valuation methods has been challenging. I have observed that sometimes the results, as seen generally in many other human-survey results, can provide less than an exact outcome and can also be contentious. I note and briefly highlight the point about valuation here to raise it as an issue, applaud the efforts that have been done for marine ecosystems thus far, and call for much further work on the topic.

My economist colleagues remind me that socioeconomic studies are more than just about money or even valuation; they are ultimately about human behavior. I concur. Thus, as we step through and review aspects of Table 10.1, we need to note two particular areas that are specifically worth additional attention, especially as they can affect human behavior.

The first is the issue of ownership (e.g. Hanna 1998; Edwards 2003). The second is structuring a management (or other institutional) system to get the incentives right (Hanna 1998; Charles 2001; Grafton *et al.* 2006). I treat these together, as they are often quite interrelated. Fisheries are often a prominent, if not *the*, contemporary example invoked for the "tragedy of the commons" thesis (Hardin 1968; Ostrom *et al.* 1999). If no one owns the resource, or if everyone owns the resource (which functionally end up looking similar), then the concern and care for the long-term viability of the resource are trumped by the short-term incentive to "get while the getting is good" before someone else gets it before you. Certainly a correct balance between hyperregulation and a free-for-all market is warranted to provide a modicum of reasonable competition and economic efficiency. But very simply, we need to change the "gold rush" mentality of harvesting LMRs.

I assert that to do so, dispersing ownership of these natural resources to more specifically identifiable entities will (and in the cases where this has been done, does [Scott 1989; Christy 1996; Hanna 1998; Edwards 2003]) provide a shift in incentive leading to the enhanced management of said resources. There are many such methods and approaches to doing so (Charles 2001; Grafton *et al.* 2006), and I do not specifically advocate any of them here. There are also other methods for shifting incentives (Berkes and Folke 1998; Hanna 1998; Ostrom *et al.* 1999; Charles 2001; Grafton *et al.* 2006), and I am not expert enough to comment on them, other than to say: (1) any combination of what works should be explored, including others beyond those methods that are the most popular or most strongly advocated; (2) it may be that a combination of those approaches is more likely to succeed than just any one method; and (3) the majority of my resource economist colleagues say that the best way to shift incentives so as to avert the tragedy of the commons phenomenon is to somehow include a shift in ownership as part of the process, such that there is a more clearly delineated set of responsible parties. However it happens, the incentive shift then provides a desire by resource users and related stakeholders for longer-term stability, a more dependable (less variable) business plan, and a (more) sustainably supported LMR base (and its associated ecosystem) to ensure long-term economic viability and higher economic yields.

SUMMARY POINTS

- Economics is more than just money (it's really about human behavior).
- We need to further incorporate ecosystem considerations in discussions of socioeconomics.
- We need to further incorporate socioeconomics in discussions of EBFM.
- Knowledge of socioeconomic factors is critical to the overall EBFM process (and for that matter, any NRM).
- It is extremely valuable to have stakeholder input that accounts for broader human considerations.
- These considerations are, in large part, where different tradeoffs (regarding objectives) arise.

RESEARCH REMAINING

- Determining how we best do economic valuation of ecosystem "goods and services" and other ecosystem "existence" components.
- More applications for nailing down ecosystem valuation.
- More statistical relationships among economic drivers and social and/or biological responses.
- Addressing how we insert socioeconomic considerations into ecosystem models.
- Exploring how we insert socioeconomic considerations into the entire EBFM process.

11

Management institutions regarding EBFM

All politics is local.
Tip O'Neill, former Speaker of the US House of Representatives (attributed)

What is the purpose of a management institution? Why do we need a system of governance for the management of natural resources? I will draw upon the works of others (especially Charles 2001) to explore this theme, but in many respects, we can start to address these questions by going back to first principles.

Basically, 100% of people do not act 100% legally, ethically, and morally 100% of the time. I often raise this point in various classes, and the students are mildly shocked to think about it. Then I ask these students if they all obeyed, to the fullest degree possible, every traffic law, speed limit, and motor vehicle regulation while they were driving to class, whereupon they concede the point, and grasp the need for governance.

Even if 100% of people did behave 100% legally, ethically, and morally 100% of the time, there would still be legitimate differences in perspectives, opinions, worldviews, etc. that need to be resolved. A governance system provides a venue for such differences to be resolved. There needs to be a system for addressing such tradeoffs, and governance institutions provide a framework or process in which to do NRM (Yaffee 1996).

The need for equitability in allocation of resources, particularly those natural resources that are in the public domain, is in large part the reason to have a governance structure in place. By providing a venue to debate and discuss status and allocation, an institution for the management of natural resources seeks to minimize conflict, maximize benefit to all stakeholders, and ultimately allow for the owners of the resource to have the best stewardship possible of that resource. And because we recognize that 100% of people do not act 100% legally, ethically, and morally 100% of the time, the need to provide enforcement mechanisms also stems from a governance structure.

LEGAL BACKGROUND

> Not only was an ecosystem management approach legal, it was mandatory if all applicable laws were to be simultaneously obeyed.
>
> Jack Ward Thomas (2004), former Chief of the US Forest Service

There are over 100 pieces of legislation (i.e. laws) that impact LMR management in the USA. This does not include any international treaties and conventions that the USA is a signatory to. Table 11.1 notes some of the more common laws that apply to LMRs in the USA. Other states, countries, or conglomerates of countries have similar kinds of legislation. For example, Canada's Oceans Act, Australia's Oceans Policy, the EU's common fisheries policy, some local US states' (e.g. California and Massachusetts) Oceans Acts, the FAO's code of conduct for responsible fisheries, etc., all have at their core a legislative basis and mandate to manage LMRs. Some of these laws have explicitly adopted an EBFM. These pieces of legislation provide the basis for governance and management institutions for LMR management.

Once such institutions are enabled, what policies do they enact and what guidance do they provide for such policies? Several others have amplified on those themes regarding how governance institutions should function and perform (e.g. Costanza *et al.* 1998; Hanna 1998; FAO 1995, 2003; Sissenwine and Mace 2003; Gray 2005; Crowder *et al.* 2006; Bundy *et al.* 2008).

The examples of legislation noted above not only form governance institutions, but also serve as the legislative mandates that specific policies are based on. There is then a need to "unpack" laws and policies to broad goals, then by specific goals (i.e. operational objectives; Jamieson *et al.* 2001; O'Boyle *et al.* 2005; O'Boyle and Jamieson 2006). Unpacking the laws and platitudes is conceptually a simple exercise, but in practice is often challenging. The aim is to go from national (or international) laws and policies (e.g. Endangered Species Act, ESA) to conceptual objective(s) (e.g. biodiversity is good), then ultimately to operational objective(s) (e.g. no loss of species richness). See O'Boyle *et al.* (2005) and O'Boyle and Jamieson (2006) for further discussion and examples (Table 11.2).

Table 11.1. *Significant US legislation influencing living marine resources and marine ecosystems*

1900	Lacey Act
1929	Migratory Bird Conservation Act
1934	Fish and Wildlife Coordination Act (FWCA)
1970	National Environmental Policy Act (NEPA)
1972	Coastal Zone Management Act (CZMA)
1972	Clean Water Act (CWA)
1972	Marine Mammal Protection Act (MMPA) (1994)
1972	Marine Protection, Research and Sanctuaries Act
1973	Endangered Species Act (ESA)
1976	Magnuson–Stevens Fisheries Conservation and Management Act (MSFCMA) (updated in 2007)
1980	Comprehensive Environmental Response, Compensation and Liability Act (CERCLA or "Superfund")
1986	Interjurisdictional Fisheries Act (IFA)

Also, but not limited to: Anadromous Fish Conservation Act; Atlantic Striped Bass Conservation Act; Atlantic Coastal Fisheries Cooperative Management Act; Atlantic Tunas Convention Act; Northwest Atlantic Fisheries Convention Act; Central, Western, and South Pacific Fisheries Development Act; etc., plus all the international treaties and agreements that the US is a signatory to.

Table 11.2. *Example of unpacking laws and policies to conceptual objectives, and on to operational objectives*

Conceptual objectives		Operational objectives
Objective	Maintain productivity	Consists of a verb, indicator, and reference point, e.g.:
→Subobjective	→Trophic transfers	
→. . .	→Forage species	
	→Target escapement	
	→(Maintain) biomass	Maintain biomass of forage species >50 000 t

Source: Adapted from O'Boyle and Jamieson (2006).

The point is that once laws are unpacked and objectives stated, there is a high potential for there to be "orthogonal legislation." That is, it is entirely conceivable that the objectives from some laws might be incompatible with those from other laws. For example, which law trumps the utilization of LMRs in a region? Do we pursue the Magnuson–Stevens

Act and maximize fishing, or do we pursue the MMPA or ESA to minimize any deleterious effects of fishing on the forage food of a marine mammal species? Do we temporarily halt all ocean-use activities until we can ascertain their cumulative impacts as per NEPA, or not worry about utilizing LMRs because a body of water could be eutrophied or otherwise polluted to the point of needing remediation under the CZMA or CERCLA?

Thus, there is a need for management institutions to sort out, or at least coordinate, the tradeoffs among all the objectives of said laws. One can readily see that even if there were no biological, ecological, or oceanographic tradeoffs driving the need for EBFM, there certainly would be legislatively driven rationales for conducting EBFM. And without an institution to do so, it could become chaotic. Even with institutions to do so, there is a clear need to ensure that all laws are met. The only feasible way to do so is to try ecosystem-based management.

GOVERNANCE AND MANAGEMENT INSTITUTIONS

We have noted that governance and management institutions provide the framework or process in which to do NRM. Recall the three pillars of any management system from Chapter 4. The first and last steps in that process need a place for such discussions to occur. As we have just noted, several laws allow societies to legally develop and implement the "institution" of this management system. This provides for a "place" for all interested parties to have a voice or otherwise express their opinion on the management of a LMR. As we have discussed, such NRM governance institutions also provide a venue for the process of "unpacking" national laws to develop and apply specific policies.

But most importantly, as they become codified, these institutions ultimately form an agreed-upon basis from which management decisions are made. They lay out the process, protocols, procedures, timing, constraints, etc., from which decisions are made. One does not necessarily have to like particular decisions, but the institutional structure at least allows one to know how such decisions are made.

These institutions particularly address the need for equitability in allocation of resources, particularly those natural resources that are in the public domain. In the USA there are national laws that regulate and call for stakeholder participation (e.g. the Federal Advisory Committee Act [FACA], the Freedom of Information Act [FOIA], Government in the Sunshine Act ["Sunshine Act"]), which must be applied to these NRM institutions. Additionally, there are other US national laws that exist to avoid the development of monopolies for areas of public interest (e.g. antitrust, public goods regulation, etc.; the Sherman Antitrust Act, the Clayton Act, the Federal Trade Commission Act). Examples include laws and resultant policies that have affected phone companies, public utilities, and similar industries in recent years. Basically, in the USA there is a very strong principle of participatory and democratic decision making, and that is largely true in much of the developed world. It is generally a good idea to apply this principle to the management of LMRs for some good reasons.

The perception of unfair allocation, particularly of a public resource, is a no-brainer for initiating a law suit. From a scientist's, fisher's or manager's perspective, a lawsuit equals increases in cost, energy, time, penalties, bad press, etc.; it is most normally a no-win situation. A lot of socioeconomics factors play off one another even in a well-designed governance institution. In many respects, that is the idea. The good institutions allow for this and still engender agreement. The bad ones host this playing off of multiple factors to the point of becoming mired in mud, resulting in poor or limited decisions. If such institutions continue or are perceived to be inequitable in their resolution, then lawsuits abound.

Additionally, the need exists to demonstrate that all "reasonable and rational" factors/opinions/perspectives have been considered in a decision-making process (basic premise of most laws; e.g. in NEPA, CERCLA, FOIA, etc.). How one defines "rational" can be contentious, but most reasonable individuals can generally agree on the basics before allocation negotiations begin. Ultimately, there needs to be a "place" where resources are divided up so that the fewest number of people are upset. That is the purpose of these institutions; to provide both the "place" and the process for allocating resources. By not being participatory or relatively democratic, the allocation process runs the risk of being viewed as fatally flawed.

Governance and management institutions also form the basis for accountability of how the resources are utilized. These institutions provide the requisite enforcement mechanisms. For example, once the specific strategic and tactical advice for LMR in an ecosystem has been established, what are the consequences of violating management decisions? And similarly, but from a systemic or institutional perspective, what are the penalties for delayed management action and who, ultimately, is accountable for the management of LMR? By not being participatory or relatively democratic, the buy-in from stakeholders is typically much less, leading to increased challenges in enforcement and lack of trust that accountability for wise stewardship of the resource will be enacted.

Thus, in these institutions there need to be mechanisms emplaced for involving stakeholders. In the USA, there is now certainly representation at the Council, Commission, or similar organizational levels, along with many of the subcommittees associated with these institutions. There are a lot of local "special interest" groups and organizations that also serve as a venue to provide a voice, but those are often at different spatial (or political boundary) scales than the more regional fisheries management bodies. It's worth asking whether there are more ways to involve stakeholders, and this question applies in both the USA (with which I am most familiar) and elsewhere.

Beyond "a voice at the table" where formal decisions are made, there are a few other mechanisms for involving stakeholders that have shown some merit. Some include a range of workshops, both training and interactive, whereby fishers and scientists and managers effectively all get together and discuss issues in a more informal environment. An excellent example that I have been involved with is the Marine Resource Educational Partnership (MREP, a program based out of the Gulf of Maine Research Institute [GMRI] and jointly funded by several parties associated with marine fisheries). Many of the interactions I have had with stakeholders at MREP are ones that would never occur at a more formal council or commission meeting. As such, many of those interactions have headed off numerous potential conflicts by simply clarifying misperceptions, misunderstandings, or miscommunications among all the parties involved. Similarly, the increasing role of cooperative research has shown related positive outcomes with the added benefit of increasing the knowledge base from which we make management decisions. Additionally, there is always the need for enhanced involvement in various technical committees, plan teams, etc., where the specific details are ironed out. The Australian example of this (Smith *et al.* 1999, 2008) is particularly informative as it has fisher, manager, NGO, and scientific representation at all steps in the fisheries management advice-producing process. The point of all this, as highlighted in the last chapter, is the need to interact on a personal level.

A final observation: in order to mitigate much of the frustration I have seen expressed by all involved in the process, it may be necessary to remove complexity from the management institutions and their related processes.

COMBATING COMPLEXILLINESS

Keep it simple, stupid.

Anonymous

Kindly allow me to introduce a term I have fashioned: complexilliness. When you have something with both complexity and silliness, it has complexilliness. I will illustrate the concept with an example that came up when trying to parameterize a full system model (cf. Chapter 8).

In the Northeast United States (NEUS) region, as I and colleagues attempted to fit an ATLANTIS model, we needed to define all of our fisheries. Using a variety of sources, including type of gear (139 winnowed down to 14); vessel sizes (5 tonnage classes plus 2 smaller types); main targeted species groups (22 groups winnowed down to 7); and number of ports (77); we came up with an original definition of 52 822 distinct fleets. Thus, we redefined fleet as an amalgamated gear × target species/group combo to obtain 16 main fleets that minimized different ports, vessel sizes, etc. The resultant actual gear–species interactions, including bycatch, was then a possible 182.

From the revised definition of the NEUS fisheries being composed of the 16 main fleets, we wanted to know how many management plans and their associated management measures would then apply to each fleet. The number of management plans in the entire NEUS region, including ESA/MMPA and NEPA-like considerations, was 32. By omitting those singularly coastal plans for species such as horseshoe crabs, we reduced the number of salient plans we had to consider to 25. This resulted in 400 possible interactions between fisheries (i.e. fleets) and plans to consider, including bycatch.

When examining this range of plans, the average number of management tools or measures (see Table 11.3 for examples of common fisheries management tools) used per plan was between five and six (with no plan having fewer than three management measures). Remember that there are 400 possible fishery–plan interactions to consider, including bycatch. Thus, there are approximately 2000 to 2400 fishery–plan alternatives to track, as these management measures can affect the execution of the fisheries. For perspective, there are 1562 species interactions in the US Northwest Atlantic food web (Figure 11.1; Link 2002a).

I have seen this tendency the world over. In effect, one almost has to be a lawyer to go fishing in such situations so as to avoid any unintentional misappropriation of harvest, gear, or plan measure. It's no wonder that fishers and other stakeholders become frustrated.

Table 11.3. *Typical fisheries management tools or measures*

Gear modification	bycatch reduction devices
	turtle excluder devices
	mesh size/hook size
	rollers
	head/foot rope modifications
	net opening size
Catch controls	total allowable catch and other quotas
	multitier quotas
	non-target species, protected species, and habitat caps
	bycatch quotas
Effort controls	days at sea
	number of permits
	number of vessels
Technical controls	vessel size
	engine horsepower
	vessel monitoring system (VMS)
	crew size limits
Temporal management	rolling closures
	spawning closures, foraging closures
	other seasonal closures
Access and property rights	individual transferable quotas/individual fishing quotas
	community based
	comanagement
	cooperatives
Economic incentives	taxes and tax breaks
	subsidies
	reporting
	buyback programs
	compliance incentives/rewards
Spatial management and zoning	closed areas
	marine protected areas
	reserves
	ocean use zoning
Social incentives	reeducation (i.e. job training in another sector)
	cultural shifts

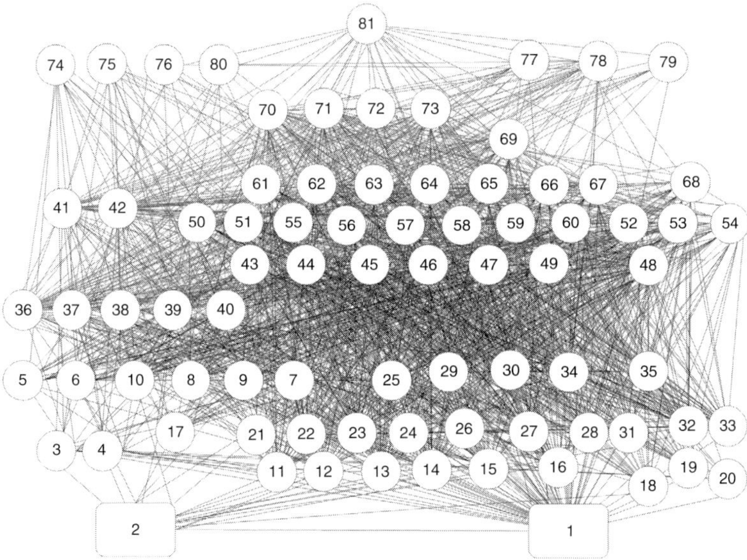

Figure 11.1 The NEUS food web diagram. (Adapted from Link 2002a.) Compared to the potential 2000–2400 Plan–Fishery interactions in the NEUS, this realized web of interactions is actually less complicated, highlighting the complexilliness of the current management process when all species and all plans and all management measures are considered separately.

This example serves to illustrate the point that management systems (not limited to only those in the situation just noted), and more broadly, governance institutions, typically tend towards two things (and have categorically had these tendencies):

- highly complex flow diagrams of process and procedure that become unwieldy
- an increasing number of rules and regulations that become unwieldy.

That is, they usually become complexilly. There are probably specific socioeconomic phrases of which I am unaware in the fields of organizational dynamics or industrial psychology for organizations to tend towards complexity and inefficiency. Yet very simply, such situations can ultimately end up working at cross-purposes to the original reason an institution was established in the first place. This also speaks to the culture of an organization, and how the culture (habits, procedure) often simply just gets handed down, without any understanding of why it is, other than "It's just the way we do it." It is time to reevaluate how we do fisheries management, and I assert that implementing an EBFM may actually be an opportune time to do so.

I certainly want to avoid falling down the other slope of parsimony and end up suggesting something that is too simplistic for the EBFM task at hand. But I trust that I have demonstrated (and that one will concur) that we are in much less danger of being overly simplistic than of violating the other side of parsimony with what are already too burdensome and complicated approaches to NRM of LMRs.

Moreover, I strongly assert that EBFM actually has the potential to make management institutions simpler. Any intelligent reader will ask how, given all the complexities of an ecosystem. I concur that ecosystems are complex, but that does not mean that our management institutions need be equally complex. Certainly the ecosystem might be more complex and less certain than solely a stock focus, but with an EBFM we are considering a broad range of factors *all at once*! That is, rather than piecemealing all of our management processes to the point that they become unwieldy, we need to do a better job of coordinating among those pieces and making concurrent management decisions. EBFM by definition forces us to do so.

As an example, recall the two-tier quota (i.e. total allowable catch) approach from ICNAF (Brown *et al.* 1976) or the North Pacific Fishery Management Council from Chapters 5 and 8 (Witherell *et al.* 2000). What if we: (1) tracked SS to ensure that none were overfished; (2) checked their status routinely, but did formal SS assessments a lot less frequently; and (3) managed total fishery harvest in the system based on an aggregate biomass group that exhibited more long-term stability and sustainability than the individual stocks? Can you envision how that would simplify the situation? Instead of 25 plans, all requiring compliance with a myriad of management tools, this approach would require one plan with a coordinated set of (much fewer) management measures. I think that ultimately this is the fishery ecosystem plan (FEP) that the EPAP (NMFS 1999) and both Commissions (POC 2004; USCOP 2004) had envisioned.

INSTITUTIONAL IMPLEMENTATION

> The buck stops here.
>
> Idiom, popularised by Harry Truman

I previously alluded to the idea that governance and management institutions rely heavily on THE PLAN for managing LMRs in an ecosystem. There are numerous fishery management plans (FMPs) today; the challenge is to develop fishery *ecosystem* plans (FEPs). The debate about FEPs has been twofold: (1) what they should contain; and (2) how much "teeth" they should have (i.e. how much authority they carry and convey). Regarding the first question, there needs to be as much information in a FEP as is necessary to make and inform ecosystem-based LMR management decisions, but no more. That is, one probably does not want a 20-page recapitulation of SS reports, nor a 400–500+ page ecosystem status document (e.g. Link and Brodziak 2002; Boldt 2008).

As for the second concern, the debate is ongoing whether a FEP:

- is an umbrella document identifying all the salient issues without making enforceable management recommendations (e.g. Ragnarsson *et al.* 2003; CBFEAP 2006)
- is a coordinating document pointing to other species-specific or sector-specific plans that contain management recommendations (e.g. EFEP 2004; NPFMC 2008)
- sets the actual tactical and strategic management measures for an entire ecosystem and its component species and fisheries in one fell swoop.

As there is yet no mandate in the USA for FEPs, as in some other countries, the role of THE PLAN has been fluid, and will continue to be. As the discussion regarding the role of FEPs continues to develop, I will go on record as saying that either option 2 or 3 is preferable to the first regarding the role and "bite" of FEPs.

There remains a conundrum with respect to EBFM and institutions. Do we work within existing structures or develop new ones? My experience and observations are that most countries are trying to proceed with EBFM within existing structures, as an evolving process. The countries that have new management institutions (or at least novel structures, processes, content, and output) for EBFM also have new enabling legislation.

In some countries, fisheries management costs are capped at a fraction of the value of the fisheries (e.g. Smith 1998; A. D. M Smith, CSIRO, personal communication, 2004; John Annala, GMRI, personal communication, 2008). Doing so has its pros and cons, but moving towards some form of cost-capping relative to the value of the fishery might make the institutional costs of management (which are much more than any dollar amount, but also the efficiencies of the process, the straightforwardness of the process, the timeliness of the process, the level of complexilliness of the process, etc.) much more palatable and less inclined towards complexilliness. But this would need to be balanced against needed costs associated with non-revenue-producing ocean uses and management (e.g. PETS). Again, the specifics of this are open for debate and consideration as we evaluate how much energy and effort we spend to produce what have been viewed as less than fully effective decisions for LMR management.

At the end of the day, we need to ask ourselves why there has generally been a failure to make decisions for LMR management issues. Rosenberg *et al.* (1993), Beverton (1998), Caddy (1999), Caddy and Cochrane (2001), Sissenwine and Mace (2003), and Caddy and Agnew (2004), among others, provide insightful commentary and reviews on the matter. The conclusion of their observations is that there is usually some combination of "end-arounds" to an existing process, fear for not acting or for acting in an irrational manner, and, although not stated directly as such, some form of complexilliness.

I submit that as we move towards implementing EBFM, we need to examine the management institutions we have and evaluate their effectiveness as rigorously (sensu Smith and Link 2005) as we do the stock and ecosystem assessment part of the three-tier management process. We want to avoid the extremes of changing the structure too often or becoming so entrenched in how we do business such that it never changes. We simply need to be open in considering if there are ways in which we could improve our management institutions; improvements to ensure that we address all the requisite tradeoffs facing us and to streamline our management process.

SUMMARY POINTS

- There are over 100 laws that impact LMR management in the USA, and there are similar pieces of enabling legislation elsewhere around the world.
- Legislative mandates (i.e. laws) form the basis for LMR policies, which form the basis for conceptual objectives, which form the basis for operational objectives.
- Management institutions should avoid "complexilliness."
- There is a need for a management institution to sort out the tradeoffs among all the objectives of said laws.
- Management institutions need to set up mechanisms to include a vast array of stakeholders.

RESEARCH REMAINING

- An examination of the best balance of social, economic, and conservation/biological objectives such that all applicable laws are met in a particular situation.
- Case studies and "autopsies" of what worked and what didn't globally, with respect to governance and management institutions.
- Development (and application) of governance best practices.
- Further identification and refinement of those issues that can hinder the implementation of any management institution.

12

It's all about tradeoffs

A pessimist is one who, when he has the choice of two evils, chooses both.
Oscar Wilde (attributed)

We've now reached the ultimate chapter of this book, and if you have stayed with me on this journey thus far, you're probably asking how I will tie it all together. Allow me to do so by returning to the catchphrase of this book: tradeoffs.

In this book, we've discussed the underlying philosophical and contextual basis for prosecuting fisheries and enacting EBFM. We have explored the underlying biology and related natural sciences that form the basis of our understanding to enact EBFM. We have considered the socioeconomic considerations critical to doing EBFM, and we've even noted the need for an institutional framework to address EBFM issues. But, what does that process look like, and ultimately, what are we trying to obtain as the outcome of such a process? How do we explore, examine and address the tradeoffs inherent to such a wide range of considerations?

I submit to you that without a place, process, or some such mechanism to sort out the social, economic, biological, ecological, and cultural tradeoffs, we will remain mired in the same difficulties we have faced for centuries.

TRADEOFFS: WHAT DO WE WANT?

. . . the biological objective of ecosystem management must specify the species mix that is desired. . .

Peter Larkin 1996

Recall the three pillars of any management system (Figure 4.1, reproduced here as Figure 12.1). In that context, the first step is goal

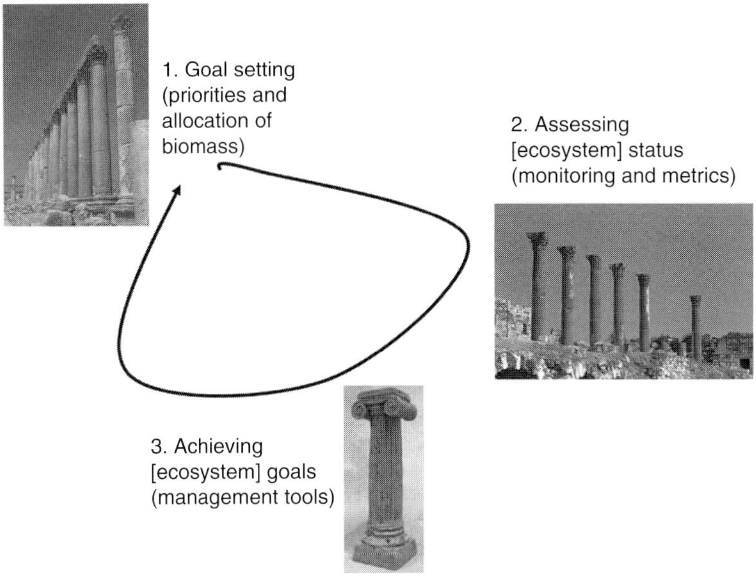

Figure 12.1 A common three-step decision theoretic. These represent "pillars" of any management system. Here, as compared with Figure 4.1, the iterative loop *à la* MSE has been denoted.

setting or, in particular, determining the desired prioritization and allocation of biomass in an ecosystem. Doing so is the ultimate in dealing with tradeoffs. What this effectively implies is that any NRM process needs to specifically address allocation issues. Allocation of LMRs is what EBFM is all about. But from an ecosystem perspective, there are things we need to account for first before we allocate LMRs into harvestable sectors.

Thus, EBFM goals should be presented as such to facilitate a framework within which to evaluate a prioritization among a wide array of tradeoffs. Recall from Chapter 3 the widely shared underpinnings of goals, with an implied ordering, which resulted in the following generic prioritization:

> Economic objectives shall be subordinate to social objectives, which shall be subordinate to biological objectives, which shall be subordinate to human existence.

This has provided a clue to help sort out orthogonal legislation. And although specific, those are not yet fully unpacked (Chapter 11) into operational objectives. Thus, we still need a biological basis from which to depart.

Table 12.1. *Steps for implementing EBFM*

Steps
Determine ecosystem productivity (S, H)
Protected and sensitive species tax (S, T)
Calculate total biological cap (S, T)
Allocate total biological cap into sub-caps across species or species groups (S, T)
Allocate sub-caps to different fleet sectors
Establish management measures that best allow sectors to sustainably achieve their caps
Enforcement and accountability

S, strategic, H, heuristic, T, tactical levels of management advice and information.

I submit to you that following the broad prioritization just given, and accounting for the production of an ecosystem, there is then a logical sequence of goals for implementing EBFM to the point that tradeoffs are addressed and each harvestable sector is judiciously and equitably allocated (Table 12.1). Examples of these goals include: (1) resource removals (i.e. harvest) shall not exceed the productive capacity of an ecosystem; (2) adequate biomass, production, or energy shall be maintained to ensure that protected species, non-target sensitive species, and essential fish habitat (and other critical ecosystem components and functions) are conserved before harvest levels are determined; and (3) no species shall go extinct.

Or, more directly tied to national or international legislation, the goals may be stated as: (1) prevent overfishing; (2) protect sensitive species; (3) conserve biodiversity (including species, genetic and functional); (4) protect critical habitat; (5) maintain trophic resilience; and (6) maximize societal benefits subject to the previous goals. And so on, with the particular goals being specific for a system, but generally covering the main elements as noted in appropriate guidelines (e.g. Restrepo 1999; FAO 2003).

Further, a second tier of goals would generally specify which species (or species groups) all stakeholders would prioritize in a system, subject to constraints and thresholds of some of the issues discussed above. Examples of the second type of goals might be: maximizing (ecologically constrained) yield of gadids over pleuronectids; reducing effort on small pelagics to enhance billfish production and yield; and so on. Systemic productivity, with set-asides subtracted for PS, NTS, and habitat, should be the basis for setting these harvestable biomass goals.

This is akin to turning Schaefer's (1956) "levels of investigation" tier system upside down. It also recognizes the interconnectedness of marine ecosystems and the lower trophic level production required to produce TS. Objectives and goals to set specific harvesting alternatives should be evaluated within ecological and economic feasibility (Edwards *et al.* 2004).

Stated another way, once we know how productive an ecosystem is (i.e. how much it can produce based upon readily available measures of primary production), we can then allocate that production into biomass set-asides for mandated species or regions of protection and then into harvestable sectors. It is in the allocation of harvestable sectors that there is plenty of room for discussion and creative thinking. And in large part, that is where the quantitative outputs (see Chapters 6–9) come into the process. Yet too often, the debate and creative thinking has focused on other aspects of this approach. That is, we debate about setting goals over which we have no basis or right to be debating.

Certainly, as alluded to above, there needs to be room for negotiation in the management of fisheries. What I am proposing is that we set total systemic caps or limits based upon an assessment of the productive capacity of the system, followed by set-asides for PS, NTS, and habitat considerations (e.g. sufficient forage base for protected marine mammals). Once the set-asides have been accounted for and subtracted from the systemic cap, harvestable biomass can be calculated. Negotiations for harvestable biomass removals among species groups (not to exceed SS overfishing thresholds per species) are constrained by ecology, and the goals and targets set for them need to account for these factors. Once harvestable biomass for species groups has been set, negotiations among fisheries sectors need to occur. The entire process needs to not only be cognizant of market fungibility (Solow 1991; Edwards *et al.* 2004), but planning for the economic considerations should be given prominent discussions in the negotiating phase of the process. Formal evaluation of alternatives needs to be explored in an interactive manner to ensure that choices are feasible and meet inherent ecological constraints (Smith 1998; Sainsbury *et al.* 2000; Sainsbury and Sumaila 2003; Smith *et al.* 2008; Levin *et al.* 2009).

This sequence of events leading to the allocation of LMRs for harvest needs to explicitly recognize that within EBFM goals, there will be both nonnegotiable and negotiable items. Too often we have tried to negotiate on items that are simply not negotiable from a straightforward, scientific basis, or should not be negotiable from a legal basis.

Both of the Ocean Commissions consistently call for some items that are simply not debatable every time they are presented (POC 2004; USCOP 2004). We need to clearly note those items that are nonnegotiable (i.e. what we do not mess with) from those items that are indeed negotiable (i.e. things that we can tweak and fiddle with).

The nonnegotiable items for EBFM should include:

- The implementation of management strategies (using a suite of possible management tools, measures, and approaches) to achieve these objectives shall minimize complexilliness.
- The level of biological caps (and related limits) shall not be negotiable.
- The conservation objectives shall be addressed prior to harvesting allocations (e.g. set-asides of biomass, forage, or habitat for those species, groups of species, or regions specifically identified as protected, endangered or threatened by law).

That then leaves a wide array of things that are negotiable for EBFM, loosely grouped into these two main considerations:

- Allocation among harvested resources should be done to account for the species mix that provides the most acceptable range of options with respect to fish biomass production and economic value, without eroding the sustainability of any given stock or ecosystem component productivity. That is, allocation among harvested resources should be executed among species with respect to ecosystem constraints as noted above.
- Allocation of allowable harvested species should be executed among fishery sectors in an equitable, safe, and feasible manner.

By establishing the process that accounts for these EBFM goals in a sequential, prioritized manner, we allow ourselves to address broader ecosystem-level considerations, handle potentially orthogonal legislation, and very clearly note what limited amount of LMR productivity we have to work with and thereby allocate for harvestable use. This sequential process allows us to address a range of tradeoffs in a reasonable and logical order. Doing so provides order to what is otherwise a complex situation (legally and biologically). More so, this process explicitly recognizes that there are inherent limits as to the productivity of any ecosystem, and thus what we are able to harvest. Specifically and explicitly recognizing the limits of ecosystem productivity actually forces us to focus on the allocation of harvestable biomass, which are the tradeoffs that we can address. At that point, much of our quantitative and analytical machinery can then be set into motion.

PROCESSES THAT DEAL WITH TRADEOFFS

So now that we've established a generic prioritization and identified a process to address tradeoffs, how do we evaluate any resultant proposed allocation schemes?

Many places are now moving towards a MSE process (management strategy evaluation, or management procedures, MP; e.g. de la Mare 1986, 1998; Butterworth *et al.* 1997; Butterworth and Punt 1999; Smith *et al.* 1999; Sainsbury *et al.* 2000; Schnute and Haigh 2006; Butterworth 2007; Schnute *et al.* 2007) as part of the NRM institutions that interface between the ecosystem/biota and socioeconomic considerations (Yaffee 1996). Figure 12.2 shows a schematic of a MSE process, which is effectively the same as the three pillars of any management system repeatedly referenced in this book (Figure 12.1). The point is that a MSE is cyclical and provides feedback for each step in the management process, allowing information to flow from one step, so as to inform another.

Management strategy evaluation is based on the Adaptive Management (Walters 1986) concept and has been used successfully in CCAMLR, South Africa, and Australia (e.g. de la Mare 1986; Butterworth *et al.* 1997; Butterworth and Punt 1999; Smith *et al.* 1999; Sainsbury *et al.* 2000; Dichmont *et al.* 2006a, 2006b; Butterworth 2007) for classical single species, multiple species, and now multiple sector ecosystem applications. It is not without its detractors (e.g. Rochet and Rice 2009), but has proven to be fairly flexible and adaptive, even regarding any of its criticisms. The MSE process is conceptually simple, but the process can, in practice, become challenging given all the considerations involved with EBFM. The actual process, however, is much simpler than piecemealing everything together species by species in the common stock assessment approach.

So what does management strategy evaluation do? The aim of MSE is to examine tradeoffs in performance of alternative management strategies (i.e. combinations of various management measures at different levels; cf. Table 11.3) across a range of management objectives, and in so doing, to provide a basis for selecting a management strategy (or range thereof) that will facilitate meeting management goals and

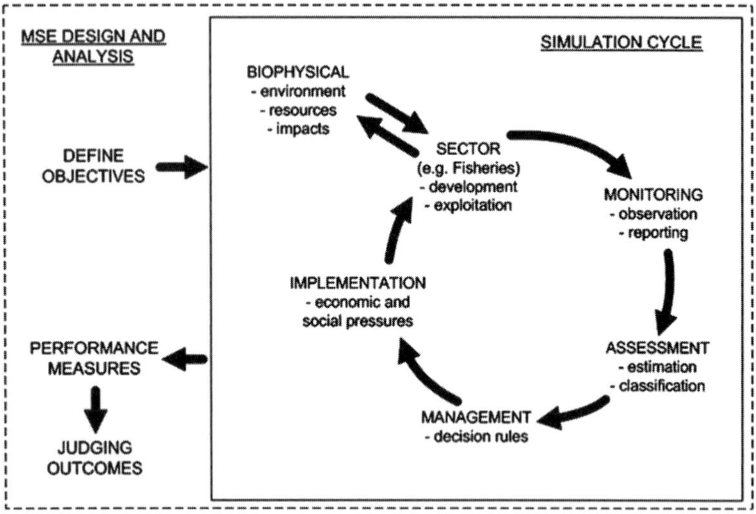

Figure 12.2 A typical schematic of management strategy evaluation (MSE). (Adapted from Fulton *et al.* 2007; see also de la Mare 1986, Butterworth *et al.* 1997; Smith *et al.* 1999, Sainsbury *et al.* 2000, and Schnute *et al.* 2007.)

objectives. This means the ability to explore allocation among species and across sectors. Key ingredients for successful MSE include:

- a clearly defined set of management objectives
- performance measures related to each objective
- a set of alternative management strategies
- a method for evaluating the likely performance of each strategy.

Basically, the rationale for doing MSE is that it provides a better chance of achieving management objectives (by pre-testing to identify robust strategies) and provides greater certainty for all stakeholders (i.e. agreed rules for decision making) than other, albeit similar, processes. MSE is designed to achieve a pre-agreed-upon balance between competing management objectives, to be robust to current scientific uncertainty, and to demonstrate to the community that one is managing responsibly (meeting the precautionary principle). Again, doing so allows for the specific evaluation of allocation among biomass and fleets.

The process as generally described has within it an "operating" model that could be one of any of the models or approaches from Chapters 7–9. That operating model is used to simulate various outcomes based upon the range of management strategies to identify those (strategies) that are most likely to result in successful management.

In other words, when you simulate a virtual ocean (or parts thereof) and the model becomes really messed up, then: (1) you know what caused it, (2) you didn't actually mess it up, (3) you can learn from it, and (4) no one is upset since it happened on a computer, not the real world. Utilizing such an approach has some obvious benefits.

Variants of a MSE approach (beyond the semantics of MP) include forms of risk assessment or risk analysis (RA; e.g. Peterman 1990, 2004; Francis 1992; Lane and Stephenson 1998; Fletcher 2005; Fletcher *et al.* 2005; Chapter 9) and integrated ecosystem assessments (IEAs; e.g. Ayensu *et al.* 1999; Levin *et al.* 2009). The thing all these processes/frameworks have in common is that they evaluate the probability of successful management before said management is enacted, and they provide iterative feedback to improve the NRM of LMRs. All this is done such that management objectives are most likely to be met on a satisfactory basis. The caveats between these different approaches are less critical than actually just having such a process in place as part of the management institution. The ability to explore among a range of management options, strategies, or alternatives is the best way to address tradeoffs among those EBFM goals that are negotiable.

Using MSE has helped to flesh out viable management objectives that are both operational and feasible. Again, the point of MSE is to examine tradeoffs in performance of alternative management strategies across a range of management objectives. MSE is not designed to find an optimal solution. Further, MSE is designed to help decision makers, not prescribe outcomes. As such, it can incorporate a wide range of objectives (including ecological and economic) and evaluates the risk of not achieving a desired state, or of achieving an undesired state.

RETURNING TO MANAGEMENT OPERATIONAL OBJECTIVES

It is valuable to use the MSE process to help codify the prioritized list of objectives at the beginning of this chapter (see also Table 3.1). In so doing, MSE (or a process like it) should also be used to specify the most acceptable mix of species, and thus harvestable biomass, across the fleet. But it should also do more than explore desirable states as this series of tradeoffs is evaluated. That is, MSE (and its operating models) should not only provide strategic information; the process (not necessarily the same models) should also be used to establish tactical advice.

Recall the "unpacking" discussion from Chapter 11. The need for operational objectives is also critical for EBFM. These operational objectives identify specific reference points/targets/limits/thresholds for a series of indicators, at both the single species and ecosystem level. In MSE and simulations using its operating models, there is then a pre-agreed set of "control rules" or triggers if the thresholds are exceeded, such that management measures are then enacted, virtually (see review of control roles by Deroba and Bence 2008; see de la Mare 2005; Utne 2006). Recall from Chapters 6–9 some of the SS, MS, and ecosystem indicators (and methods to produce them) that could be useful for serving as operational objectives. Remember also how one would establish thresholds associated with such indicators. From Chapter 11, recall that there are standard fisheries management measures and tools, and once the operational objectives are defined, there is a wide range of measures/tools/options (Table 11.3) with which to enact them. For EBFM, these are the same as standard FM tools, just applied in clever or unique combinations.

Let's work through an example of a groundfish fleet and pelagic fleet operating in the same ecosystem. This serves as an example of a common situation where the need to address tradeoffs is particularly heightened. Let us say that after accounting for the nonnegotiables, you note general objectives to maximize economic value of groundfish harvest for one fleet, and maximize long-term yield and jobs for a small pelagic fleet. Then you could identify a set of indicators to track, such as the ratio of pelagics to demersals; biomasses of major species groups (e.g. clupeids, gadids, pleuronectiformes); trophic requirements for the groundfish based on their feeding upon both the targeted and other small pelagics; the number of vessels in each fleet; and the value of each set of landed species. Then you could establish thresholds, values below or above which would be undesirable as the operational objectives (for example): a revised B_{MSY} for a small pelagic stock that accounts for both F and M2; or pelagics to demersal ratio (Pel/Dem) to be greater than 50%; the value of groundfish landings to be above \$$x$ per pound, and so on. Then you could note a range of management options that could best meet these thresholds: for example, closing the herring spawning grounds for a short season; changing mesh size for either fleet; keeping a cap on a target species but spreading out the fishing season in which they could be caught, etc. But then what? In many ways, we do variants of this now (albeit not in a MSE context, as in this hypothetical example).

The important aspect of MSE is that it allows us to evaluate management performance. How well would our hypothetical management tools/measures/strategies perform in the example above? What would the status of the indicators noted above be relative to the reference points (or thresholds) after a set of management actions were triggered (Figure 12.3; see Deroba and Bence 2008)? How well would both the groundfish and

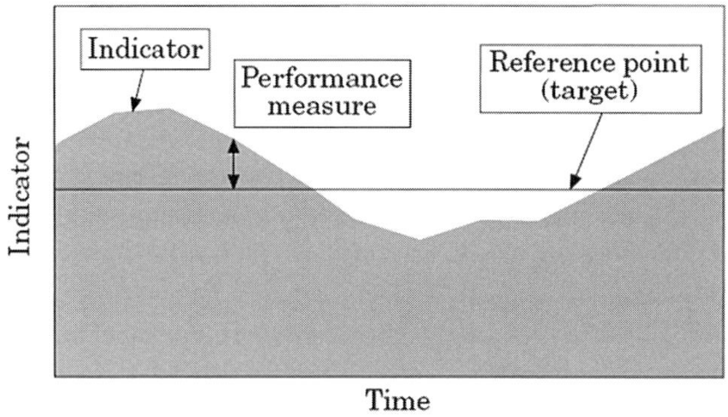

Figure 12.3 Diagram of performance measures relative to reference points (or thresholds). (Adapted from Sainsbury *et al.* 2000.)

small pelagic fleets, and the LMR they rely upon, be doing with respect to both specific, operational objectives and the broader, general objectives of use in this hypothetical ecosystem? What tradeoffs would we need to fine-tune further to better manage for all aspects of this situation?

The point is that we need to evaluate the performance of our management (sensu Walters 1986), and the virtual world of MSE provides a place for us to do that as part of the overall process, before we ever implement such management in the real ocean. If a management measure or tool is not performing well, it makes sense to evaluate it and change strategies to something more appropriate or efficacious. The key to MSE is "closing the loop" in the management cycle in order to get this feedback. Doing so is fairly rare in practice, unless institutions are structured and processes are utilized to do so.

Ultimately, if we can move the LMR NRM system to one more akin to engineering (i.e. with controls, triggers, and processes that are semi-automated; see de la Mare 2005; Utne 2006) it may help make LMR NRM more effective. Doing so for EBFM, which has myriad more factors to consider, is almost imperative.

FIN

It doesn't matter if you close the barn door once the horse is out.

Idiom

Very simply, there are more uses of the ocean than just fishing (e.g. listed in Sainsbury and Sumaila 2003; Sissenwine and Mace 2003). EBFM needs to expand the decision-making process to be more inclusive of all stakeholders in it. There is a positive trend towards recommending expanded scope of fishery agencies to evaluate all effects of fishing activities on the ecosystem, yet this is an enormous task and will likely require enhanced partnerships with those organizations with the appropriate expertise. Conversely, imperative for EBFM is improving coordination among other institutions with jurisdiction over activities that can affect fishery resources.

As we move towards multiple-sector management of a vast array of ocean uses (i.e. EBM; Chapter 2), it will become increasingly important to have the fisheries sector in an ecosystem better managed and functioning as a more unified set of stakeholders. This is because fisheries will begin to compete for tradeoffs at a much broader scale than simply species allocation across fleets. I assert that doing EBFM is the best way to do so.

Marine ecosystems are complex. Managing living marine resources – including those marine ecosystems and the people that utilize them – is even more complex. Certainly there are data and information gaps within fisheries and marine ecosystem science, and we need to provide feedback to our monitoring and observation systems to address such gaps. We need to expand the multidisciplinary studies and assessments of LMRs, and we must evaluate and tweak our management institutions to better coordinate across all the facets involved with a broad range ecosystem components being considered. We can always improve the process in which we address tradeoffs, and we could certainly use more resources to make all of the above happen.

Yet, given all the things we could admittedly improve upon, we know enough and have enough *now* to begin implementing EBFM. There is no need to wait. Waiting will simply prolong the inevitable and be imprudent given all the issues facing those of us working on NRM of LMRs. Fortunately, many places are now well under way towards EBFM (Pitcher *et al.* 2009).

I trust that the material provided in this book, albeit imperfect and incomplete, is enough to demonstrate the need for and feasibility of doing EBFM. If, after 10 to 20 years, some of the things here are more commonly enacted, I will be happy and will feel that I have made a reasonable contribution. However, if nothing changes, I suspect we will have much bigger things to be worrying about.

SUMMARY POINTS

- We need to and can establish management goals.
- Said goals need to be prioritized, and a generalized first cut at prioritization is offered within this chapter.

- There are items in EBFM that are nonnegotiable and those that are negotiable; our energy needs to be spent debating about those items that are negotiable.
- A process to explore and examine all the allocation issues is needed; MSE is one suggested option.
- Certainly we need to fill in data, information, or related gaps, but we know enough to start EBFM now.

RESEARCH REMAINING

- Continued development of operational objectives.
- Virtual and pre-emptive evaluation of various strategies.
- Case studies on the uptake of MSE.
- Examination of nonnegotiables to periodically revisit allocation issues.

Glossary

abundance (N) The number of organisms in a particular place at a particular time, usually for a given population or group of populations.

allowable biological catch (ABC) The range of fisheries removals, usually in terms of mass caught, as allowable catch for a species or species group.

Atlantic Multidecadal Oscillation (AMO) A broad-scale climate indicator that shows major, long-term trends in the differences in ocean temperature for the Atlantic Ocean. The AMO is defined from the patterns of sea surface temperature variability in the North Atlantic once any linear trend has been removed.

Australia's Oceans Policy (1998) This Australian policy was released in 1998 and continues to guide the direction of the Australian Government's program in the marine environment. The policy provides national coordination and consistency for marine planning and management, while allowing for regional diversity. The policy was established with governance arrangements including: (1) a National Oceans Ministerial Board of key Australian Government Ministers; (2) a National Oceans Advisory Group of industry, community, and government stakeholders; and (3) a National Oceans Office, which was established to provide secretariat and technical support and program delivery for initiatives under the policy. By its very content, this has adopted an ecosystem approach to ocean management.

B_{MSY} The biomass at maximum sustainable yield. Often used as a reference point.

bioeconomic models A class of models that analyze biological and economic factors simultaneously. Often these are in the form of biotic growth coupled with some form of harvest and associated valuation or pricing thereof.

biogeochemical models A class of models that account for the elemental transfer of materials through chemical, biological, and geological processes.

biological reference points (BRP) Values of certain indicators, usually F and B in a fisheries context, that take into consideration both the best possible catch and conservation of the fishery resource. These BRPs are based on a wide range of models and theories, and are then set as thresholds from which fisheries management actions are then triggered. There are BRPs based on long-term projections (LP), BRPs based on values observed during a certain period of years, and BRPs based on the two previous criteria. The BRPs can be target points (TRPs), limit points (LRPs), and precautionary points (PaRPs). Common examples include: B or F_{max}, $F_{0.1}$, F_{med}, F_{MSY}, F_{PA}, with novel, ecological indicators being considered more contemporarily.

biomass (*B*) The amount of living material in a particular place at a particular time, usually for a given population or group of populations.

biophysical models A class of models that analyze how physical processes influence biological dynamics. In marine systems, this is often of the form of ocean flow and temperature fields as they affect living marine resources.

Bonn Convention on Migratory Species of Wild Animals (1979) Aims to conserve terrestrial, marine, and avian migratory species throughout their range. It is an intergovernmental treaty, under the aegis of the United Nations Environment Programme, concerned with the conservation of wildlife and habitats on a global scale.

bulk biomass models A class of models that account for the transfer, and often mass balance, of energy or biomass within an ecosystem among the various groups in that ecosystem.

bycatch The amount of harvest of organisms other than the species for which the fishing gear was set. Bycatch is also often called incidental catch. Some bycatch is kept for sale.

Canada's Oceans Act (1996) This Canadian law provides a framework for modern ocean management. The Act calls for the Minister of Fisheries and Oceans to lead and facilitate the development of a national ocean management strategy. This particularly espouses an integrated, coordinated ecosystem approach.

carrying capacity (*K*) The theoretical maximal amount at which a population, or group of populations, can be sustained in a particular ecosystem.

catch Includes but is not limited to any activity that results in killing any resource species or bringing such species on board a vessel.

catch per unit of effort (CPUE) The number of fish caught from an area over some period of time. Typically, effort is a combination of gear type, gear size, and length of time gear is used.

catchability coefficient (*q*) The level of effectiveness of a particular gear at capturing a species.

Clean Water Act (CWA) (1972) This US law is the primary federal law governing water pollution. This law established the goals of eliminating releases to water of high amounts of toxic substances, eliminating additional water pollution by 1985, and ensuring that surface waters would meet standards necessary for human sports and recreation by 1983. Major amendments were enacted in the Clean Water Act of 1977 and the Water Quality Act of 1987.

Coastal Zone Management Act (CZMA) (1972) A US law passed to encourage coastal states to develop and implement coastal zone management plans.

Comprehensive Environmental Response, Compensation, and Liability Act (CERCLA or "Superfund") (1980) This US law is designed to clean up abandoned hazardous waste sites. Superfund created the Agency for Toxic Substances and Disease Registry (ATSDR), and it provides broad federal authority to clean up releases or threatened releases of hazardous substances that may endanger public health or the environment. The law authorized the Environmental Protection Agency (EPA) to identify parties responsible for contamination of sites and compel the parties to clean up the sites. Where responsible parties cannot be found, the Agency is authorized to clean up sites itself, using a special trust fund.

Convention for the Conservation of Antarctic Marine Living Resources (CCAMLR) The major international treaty that regulates, manages, and provides the venue for addressing issues associated with marine biota in and around Antarctica.

Convention on Biological Diversity (CBD) (1992) Known informally as the Biodiversity Convention, it is an international legally binding treaty that was adopted in Rio de Janeiro in June 1992. The Convention has three main goals: (1) conservation of biological diversity (or biodiversity); (2) sustainable use of its components; and (3) fair and equitable sharing of benefits arising from genetic resources. That is, its objective is to develop national strategies for the conservation and sustainable use of biological diversity. It is often seen as a key document regarding sustainable development.

Convention on International Trade in Endangered Species (CITES) (1973) An international agreement between governments, drafted as a result of a resolution adopted in 1973 at a meeting of members of the International Union for Conservation of Nature (IUCN). Its aim is to ensure that international trade in specimens of wild animals and plants does not threaten their survival and it accords varying degrees of protection to more than 33 000 species of animals and plants.

discards The amount of catch and bycatch that is returned to the sea and not landed.

economic exclusive zone (EEZ) The area of the ocean surrounding a nation that is claimed, exclusively, by that nation for its own uses. Usually extending 200 nautical miles from the coast, but often fewer where there are geographies and borders in a multinational situation.

ecosystem approach to fisheries (EAF) Doing within-sector single species fisheries management while cognizant of the effects of other sectors on fisheries and the effects of fisheries on other sectors, and perhaps the effects of fisheries on the ecosystem (e.g. gear impacts on benthic habitats). Although mildly distinct from ecosystem-based fisheries management, the two are used interchangeably. In EAF, though fisheries are emphasized, the inclusion of marine mammals, protected species, non-target species, etc., is usually implied. Think of starting with a stock and building up to include as many factors as are appropriate.

ecosystem approach to management (EAM) Doing within-sector resource management while being cognizant of the effects of other sectors on each other. In practice, EAM is rarely distinguished from ecosystem-based management. Think of this as starting within a sector and working up to include other potential factors arising from other sectors.

ecosystem-based fisheries management (EBFM) Managing fisheries to coordinate, account for, and include all factors in a holistic, synthetic, integrated fashion. Fisheries are emphasized, but the inclusion of marine mammals, protected species, non-target species, etc., are explicitly included. Although mildly distinct from ecosystem approach to fisheries (EAF), the two are used interchangeably. Think of this as starting with an ecosystem perspective of fisheries production that addresses specific interactions among species and with the environment.

ecosystem-based management (EBM) Applying management options to a resource sector that is holistic and integrated, accounting for all key factors and processes from the primary perspective of the entire (eco)system. In marine resource circles, this is with the explicit understanding that all involved sectors are being considered and consulted. Think of this as starting with the entire ecosystem and then working down into each of the sectors as they interact among one another.

ecosystem management (EM) A catch-all term for the general management of natural resources in a holistic, sometimes amorphous manner. It is recognized that technically this is infeasible, as we are really managing human activities in any ecosystem, but it is understood to be a moniker for more broadly considered management.

ecosystem models A class of models that account for the full range of processes that can affect biota of interest. Usually these are biologically centric and include a range of human use, physical, geochemical, or associated processes as they can influence the dynamics of the biota. These types of models are noted for their inclusiveness and, hence, complexity.

effort The amount of time and fishing power used to harvest resource species. Fishing power includes gear size, boat size, and horsepower.

El Niño Southern Oscillation (ENSO) A broad-scale climate index that is a periodic change in the atmosphere and ocean of the tropical Pacific region. It is defined in the atmosphere by the sign of the pressure difference between Tahiti and Darwin, Australia, and in the ocean by warming or cooling of surface waters of the tropical central and eastern Pacific Ocean.

Endangered Species Act (ESA) (1973) This US law was designed to protect critically imperiled species from extinction as a "consequence of economic growth and development untempered by adequate concern and conservation." The Act is largely noted for its ability to authorize the determination and listing of species as endangered and threatened.

European Union's (EU's) Common Fishery Policy (CFP) (2002) The European Union's instrument for the management of fisheries and aquaculture. This policy was established to ensure sustainable exploitation of living aquatic resources. The reform introduced a precautionary approach to protect and conserve living aquatic resources, and to minimize the impact of fishing activities on marine ecosystems. It also aimed to contribute to efficient fishing activities within an economically viable and competitive fisheries and aquaculture industry, providing a fair standard of living for those who depend on fishing activities, and taking into account the interests of consumers.

extended stock assessment models (ESAMs) Population models that include another factor that is strongly suspected of influencing population dynamics. Typical examples include population models that include predator or environmental (e.g. temperature) impacts.

F_{MSY} The rate of fishing at maximum sustainable yield.

Fish and Wildlife Coordination Act (FWCA) (1934) This US law was established to protect fish and wildlife when federal actions result in the control or modification of a natural stream or body of water. The Act provides the basic authority for the involvement of the United States Fish and Wildlife Service in evaluating impacts to fish and wildlife from proposed water resource development projects. The Act authorizes the Secretaries of Agriculture and Commerce to provide assistance to federal and state agencies in order to protect and increase the supply of wildlife and wildlife resources, as well as to study the effects of domestic sewage, trade wastes, and other pollution on wildlife. This law was last amended in 1958.

fisheries management councils (FMCs) Regional bodies that are responsible for the oversight, coordination, and management of fisheries within their regions.

fishery ecosystem plan (FEP) Plan to achieve specified management goals for a regional group of fisheries. But instead of a plan per fishery or per stock, it is a coordinated plan, from an ecosystem perspective, for all such fisheries and stocks in a region.

fishery management plan (FMP) A plan to achieve specified management goals for a fishery. It includes data, analyses, and management measures for a fishery.

fishing mortality (fishing rate, F) The measurement of the rate of removal of resource species from a population due to fishing.

food web models A class of models that account for the interactions among all the biota in an ecosystem. They often have mass balance, network, or interaction strength estimates as part of the modeling process.

Food and Agriculture Organization (FAO) An international organization, part of the United Nations, that leads international efforts to defeat hunger by providing means to improve agriculture, forestry, and fisheries practices.

Food and Agriculture Organization's Code of Conduct for Responsible Fisheries (1995) This Code contains a broad set of principles and methods for developing and managing fisheries and aquaculture. A voluntary, non-binding instrument, the code is widely recognized as the global standard for settling out the aims of sustainable fisheries and aquaculture for the coming decades. It sets out principles and international standards of behavior for responsible practices, with a view to ensuring the effective conservation, management, and development of living aquatic resources, with due respect for the ecosystem and biodiversity. The Code recognizes the nutritional, economic, social, environmental, and cultural importance of fisheries, and the interests of all those concerned with the fishery sector. The Code takes into account the biological characteristics of the resources and their environment and the interests of consumers and other users. By its content, it espouses an ecosystem approach.

Graham's Law of Overfishing Graham's Law effectively states that short of any controls or intervention, the typical pattern for a fishery is to increase catch, which lowers standing stock biomass, which leads to an increase in effort, which leads to an increase in catch (but lower catch per unit of effort), and so on, the cycle iterating until a stock collapses.

Great Barrier Reef Marine Park Authority (GBRMPA) An organization that oversees the wise use of the famed coral reef in northern Australia.

habitat suitability index (HSI) An index or series of indices that model the probability that a particular organism will occupy a particular habitat, based on a plethora of abiotic considerations, as related to their suitability to that organism.

harvest The total number or mass of resource species caught and kept from an area over a period of time.

indicator A measure of something that represents another process or metric.

integrated coastal zone management (ICZM) A process for the management of the coast using an integrated approach, regarding all aspects of the coastal zone, including geographical and political boundaries, in an attempt to achieve sustainability.

integrated ecosystem assessment (IEA) A process, tool, product, and approach that seeks to provide a synthesis and quantitative analysis of information on relevant physical, chemical, ecological, and human processes of a particular ecosystem to determine that ecosystem's status in relation to specified ecosystem management objectives, as identified by target levels and thresholds for important ecosystem features.

integrated management (IM) *See* integrated coastal zone management.

Intergovernmental Oceanographic Commission (IOC) An international organization that promotes cooperation in researching and protecting the ocean. The IOC is focused on monitoring the ocean through the Global Ocean Observing System (GOOS).

Intergovernmental Panel on Climate Change (IPCC) An international organization that assesses the scientific, technical, and socioeconomic information relevant for the understanding of the risk of human-induced climate change.

Interjurisdictional Fisheries Act (IFA) (1986) This US law provides for grants by the Secretary of Commerce to States for management of interjurisdictional commercial fishery resources, usually at the state (local) or interstate (commission) level.

International Biological Program (IBP) A large-scale research effort in ecosystem ecology in the 1960s and 1970s. The IBP was an attempt to apply the methods of "big science" to ecosystem ecology and pressing environmental issues.

International Commission for the Conservation of Atlantic Tunas (ICCAT) An international intergovernmental organization responsible for the management and conservation of tunas and tuna-like species in the Atlantic Ocean and adjacent seas.

International Commission for the Northwest Atlantic Fisheries (ICNAF) Predecessor to Northwest Atlantic Fisheries Organization.

International Council for the Exploration of the Sea (ICES) An international organization that is a leading multidisciplinary scientific forum for the exchange of information and ideas on all aspects of marine sciences pertaining to the North Atlantic, including the adjacent Baltic Sea and North Sea, and for the promotion and coordination of marine research by scientists within its member nations.

International Pacific Halibut Commission (IPHC) An international organization that is dedicated to the preservation of Pacific halibut in waters off Canada and the USA.

International Whaling Commission (IWC) An international intergovernmental organization responsible for the conservation of whales and the management of whaling.

intrinsic rate of growth (r) The rate at which a population or other group of organisms increases over a particular period of time.

Jakarta Mandate (1995) The Jakarta Mandate on Marine and Coastal Biological Diversity is a global consensus on the importance of marine and coastal biological diversity and is a part of the Ministerial Statement on the implementation of the Convention on Biological Diversity (CBD), as adopted by the Conference of the Parties at its second meeting held in Jakarta, Indonesia. This was recognized as shoring up the CBD for marine and coastal ecosystems.

K See carrying capacity.

L-index of systemic overfishing An index of ecosystem overfishing based upon the trophic level of fisheries removal.

Lacey Act (1900) This US law protects both plants and wildlife by creating civil and criminal penalties for a wide array of violations. Most notably, the Act prohibits trade in wildlife, fish, and plants that have been illegally taken, possessed, transported, or sold. Thus, the Act underscores other federal, state, and foreign laws protecting wildlife by making it a separate offense to take, possess, transport, or sell wildlife that has been taken in violation of those laws. The Lacey Act was amended as of 2008, when the Food, Conservation, and Energy Act of 2008 expanded its protection to a broader range of plants and plant products.

landings The amount (in number or mass) of organisms that are caught by a fishery and that are offloaded at a dock or brought to shore.

large marine ecosystem (LME) A regional sea that has similar properties, conditions, and management, usually with obvious geographic or political boundaries.

Law of Sequential Depletion A corollary to Graham's Law. We see this in many instances, whereby the primary targeted stock collapses (i.e. can no longer support a commercial fishery) and the fleet then moves on to target another stock, and so on, ad infinitum.

LD_{50} (50% survival relative to a toxin concentration) The level of concentration of a substance at which half of a population survives (and half dies).

limit reference point (LRP) A biological reference point indicating limits of the fishery exploitation with regard to stock self-reproduction, aiming at conservation of the resource. Distinct from a target reference point.

living marine resource (LMR) Organisms that live in aquatic ecosystems, usually the ocean, that are usually of special interest for some form of human use.

M *See* natural mortality.

M1 Residual natural mortality, a subset of natural mortality not attributable to predation.

M2 Predation mortality, a subset of natural mortality attributed to predatory removals.

Magnuson–Stevens Fisheries Conservation and Management Act (MSFCMA) (1976) This US law, commonly referred to as the Magnuson–Stevens Act, is the primary law governing marine fisheries management in the USA. The Act was amended in 2006 with some additional provisions for ecosystem approaches.

management procedures (MP) *See* management strategy evaluation.

management strategy evaluation (MSE) A process that involves assessing the consequences of a range of management strategies or options and presenting the results in a way that elucidates the tradeoffs in performance across a range of management objectives. In contrast to some previous approaches to resource assessment, it does not seek to proscribe an optimal strategy or decision. Instead it seeks to provide the decision maker with the information on which to base a rational decision, given their own objectives, preferences, and attitudes to risk.

Marine Mammal Protection Act (MMPA) (1972) This US law was one of the first to call specifically for an ecosystem approach to natural resource management and conservation. The Act prohibits the taking of marine mammals, and enacts a moratorium on the import, export, and sale of any marine mammal, along with any marine mammal part or product within the US. The Law defines "take" as "the act of hunting, killing, capture, and/or harassment of any marine mammal; or, the attempt at such." The MMPA provides for enforcement of its prohibitions, and for the issuance of regulations to implement its legislative goals. It was last amended in 1995.

Marine Protection, Research, and Sanctuaries Act (1972) This US law, a.k.a. the Ocean Dumping Act, was one of several key environmental laws passed in 1972. It originally authorized the Environment Protection Agency to regulate ocean dumping of industrial waste, sewage sludge, biological agents, NBC, radioactive waste, and other wastes into the territorial waters of the USA through a permit program. A prohibition on medical waste was enacted in 1988. The Act also established marine sanctuaries.

maximum sustainable yield (MSY) The largest average catch that can be taken continuously (sustained) from a stock under average environmental conditions. This is often used as a management goal in the USA.

Migratory Bird Conservation Act (1929) This US law sought to authorize the acquisition and preservation of wetlands as waterfowl habitat. The law, however, did not provide a permanent source of money to buy and preserve the wetlands. In 1934 the Migratory Bird Hunting Stamp Act was passed. Popularly known as the Duck Stamp Act, the bill's whole purpose was to generate revenue designated for only one use: acquiring wetlands for what is now known as the National Wildlife Refuge System.

minimal realistic models (MRMs) A class of models that account for the minimal amount of additional processes or factors that capture the major dynamics of a population without attempting to model those from the full range of all possible. Examples include extended strategic assessment models and multispecies models.

multiple (ocean) use management (MUM) An approach to manage across all ocean-use sectors, accounting explicitly for tradeoffs in objectives among the wide array of stakeholders that use the ocean or a portion thereof. Usually focused for specific ecosystems or geographic areas.

multispecies (MS) Any context (modeling, assessment, management) that explicitly accounts for more than one species simultaneously.

multispecies forecast (MSFOR) The forward projecting part of multispecies virtual population analysis; a model that takes outputs from MSVPA and makes short-term forecasts of stock dynamics.

multispecies production (MSProd) A type of model that treats each stock in bulk (i.e. is not age structured) and that accounts for species interactions. Many production models are common in fisheries; the multispecies version accounts for predation or competition explicitly.

multispecies virtual population analysis (MSVPA) A model that accounts for species interactions explicitly by having a feeding submodel coupled, simultaneously, with age-structured single species submodels.

multispecies yield per recruit (MSYPR) A model that extends single-species yield per recruit by linking the fishing mortality rates for each species. This approach accounts for species being caught simultaneously by the same fishing gears.

multivariate (MV) Any situation that has more than one set of variables that need to be considered at any given time (a.k.a. real life).

National Environmental Policy Act (NEPA) (1970) A US law that established a US national policy promoting the enhancement of the environment and also established the President's Council on Environmental Quality (CEQ). The NEPA's most significant effect was to set up procedural requirements for all federal government agencies to prepare environmental assessments (EAs) and environmental impact statements (EISs). These contain statements of the environmental effects of proposed federal agency actions.

natural mortality (M) The rate of removal of resource species from a population from natural causes.

natural resources management (NRM) Any process by which any naturally occurring resource is managed, nominally for the betterment of humans and for the wisest possible use of a resource.

non-target species (NTS) Organisms that are not directly and particularly chosen for harvest or removal, but are caught nonetheless.

North Atlantic Oscillation (NAO) A broad-scale climatic index in the North Atlantic Ocean of fluctuations in the difference of atmospheric pressure at sea level between the Icelandic Low and the Azores High.

Northeast US (NEUS) A region of the northwest Atlantic off the seaboard of the eastern USA.

Northwest Atlantic Fisheries Organization (NAFO) An international intergovernmental organization with a mandate to provide scientific advice and management of fisheries in the northwestern part of the Atlantic Ocean.

nutrients-phytoplankton-zooplankton (NPZ) A type of model that tracks the dynamics of nutrients through the lower trophic levels of a food web, often as modified by physical flow fields.

ocean zoning (OZ) An approach to set aside particular places in the ocean for particular uses. A spatially explicit form of regional ecosystem management.

optimum yield (OY) The harvest level for a species that achieves the greatest overall benefits, including economic, social, and biological considerations.

overfished A stock or stock complex is considered overfished when its biomass or number of individuals falls below a minimum threshold. A rebuilding plan is required for stocks that are overfished in the USA.

overfishing Harvesting at a rate equal to or greater than that which will meet the management goal. A stock or stock complex is considered undergoing overfishing when fishing mortality rate exceeds a fishing threshold.

Pacific Decadal Oscillation (PDO) A broad-scale climatic index of Pacific climate variability that shifts phases on at least interdecadal timescale, usually about 20 to 30 years. The PDO is detected as warm or cool surface waters in the Pacific Ocean, north of 20° N.

probability of sustainability (P_{sust}) or ecosystem-based maximum sustainable catches (EMSC) Some examples of system level reference points to detect, determine, and delineate ecosystem overfishing. Takes into account primary production of an ecosystem, total catches, and at what trophic level the catches are obtained.

protected, endangered, and threatened species (PETS) Organisms that are afforded special status, usually for protection from biological extinction, by various national or international laws and treaties.

protected species (PS) *See* protected, endangered, and threatened species.

r *See* intrinsic rate of growth.

recruitment (*R*) The amount of or process of resource species being added to the exploitable stock each year due to growth and/or migration into the fishing area. For example, the number of resource species that grow to become vulnerable to the fishing gear in one year would be the recruitment to the fishable population that year. This term is also used in referring to the number of resource species from a year class reaching a certain age.

regional ecosystem management (REM) An approach to manage ecosystems as entire entities (as opposed to just stocks or single species), and across all use sectors, but not at the full geographic extent of the large marine ecosystem.

regional oceanographic models (ROMs) In the generic (not specific) sense, a class of models that seek to capture key hydrodynamic processes.

Reykjavik Conference (2001) The Reykjavik Conference on Responsible Fisheries in the Marine Ecosystem was held in Reykjavik, Iceland. The objectives of the conference were to gather and review the best available knowledge on the marine ecosystem issues, to identify means by which ecosystem considerations can be included in capture fisheries management, and to identify future challenges and relevant strategies.

Scientific Committee on Oceanic Research (SCOR) An international body promoting international cooperation in planning and conducting oceanographic research, and solving methodological and conceptual problems that hinder research.

SHIRCS An acronym meaning systemic, holistic, integrated, relative, coordinated, and simultaneous. Used to denote the need to move from single species or single sector perspectives to consider a broader range of issues or factors that can influence living marine resources.

single species (SS) The prevailing paradigm in fisheries, whereby any situation (modeling, assessment, management) is focused solely on one species, usually a stock or specific population.

spawning stock biomass (SSB) The amount of total weight of the resource species in a stock that are old enough to spawn.

stewardship The concept of managing a resource on behalf of another, usually for an owner.

stock A grouping of harvested species usually based on genetic relationship, geographic distribution, and movement patterns.

stock–recruitment relationships A statistical pattern between the number or biomass of recruits and spawning stock biomass. Can also refer to common types of fisheries models that model such a pattern.

Stockholm Conference on Human Development (1972) The United Nations Conference on the Human Environment (also known as the Stockholm Conference) was an international conference convened under United Nations auspices. It was the UN's first major conference on international environmental issues, and marked a turning point in the development of international environmental politics.

sustainable development (SD) The pattern of resource use that aims to meet human needs while preserving the environment so that these needs can be met not only in the present, but also for future generations.

system models (or full system models) *See* ecosystem models.

target reference point (TRP) Biological reference point indicating long-term objectives (or targets) for the management of a fishery, taking into consideration the best possible catch and ensuring the conservation of the stock. Distinct from limit reference point.

targeted species (TS) Organisms that are directly and particularly chosen for harvest or removal.

total allowable catch (TAC) The annual recommended catch for a species or species group. These TACs are set from the range of the acceptable biological catch.

total mortality (Z) The sum of sources of mortality. Presented as a measurement of the rate of removal of resource species from a population by all causes. Usually includes F and M additively.

tradeoffs The need to make choices from among an array of options. Central to ecosystem-based management given the myriad of objectives, stakeholders, and species groups that are involved.

United Nations Conference on Environment and Development (UNCED) (1992) Also known as the Earth Summit, took place in Rio de Janeiro, Brazil, 20 years after the United Nations Conference on the Human Environment (UNCHE) took place in Stockholm. The central focus was the question of how to relieve the global environmental system through the introduction to the paradigm of sustainable development. Provided further codification for sustainable development.

United Nations Convention on the Law of the Sea (UNCLOS, Law of the Sea) (1982) Also called the Law of the Sea Convention or the Law of the Sea Treaty, this is the international agreement that resulted from the third United Nations Conference on the Law of the Sea (UNCLOS III). The Law of the Sea Convention defines the rights and responsibilities of nations in their use of the world's oceans, establishing guidelines for businesses, the environment, and the management of marine natural resources. The Convention, concluded in 1982, replaced four 1958 treaties. Most notably this Convention established exclusive economic zones (EEZs) for each country.

United Nations Educational, Scientific, and Cultural Organization (UNESCO) A UN agency that helps promote a wide range of technical activities.

World Commission on Environment and Development (WCED, a.k.a. the Brundtland Commission) (1987) A UN commission created to address growing concern "about the accelerating deterioration of the human environment and natural resources and the consequences of that deterioration for economic and social development." In establishing the commission, the UN General Assembly recognized that environmental problems were global in nature, and determined that it was in the common interest of all nations to establish policies for sustainable development. The Commission produced a report (known commonly as the Brundtland Report) titled "Our Common Future" that codified and solidified much of the sustainable development thinking.

World Summit on Sustainable Development (2002) Took place in Johannesburg, South Africa, and was convened to discuss sustainable development by the United Nations. The Johannesburg Declaration was a main outcome, largely emphasizing an agreement to restore the world's depleted fisheries by 2015.

yield *See* harvest.

yield per recruit (YPR) An expected lifetime yield per fish recruited in the stock at a specific age. Also, a type of fisheries model that models that process.

Z See total mortality.

References

Allesina, S. and C. Bondavalli (2003). Steady state of ecosystem flow networks: a comparison between balancing procedures. *Ecological Modelling*, **165**:221–229.

Andersen, K. P. and E. Ursin (1977). A multispecies extension to the Beverton Holt theory of fishing, with accounts of phosphorous circulation and primary production. *Medd Dan Fisk Havunders, NSF*, **7**:319–345.

Anderson, E. D. (2002). *100 Years of Science Under ICES: Papers from a Symposium Held in Helsinki*. ICES Marine Science Symposia 215. Copenhagen, Denmark: ICES.

Ascher, W. (2001). Coping with complexity and organizational interests in natural resource management. *Ecosystems*, **4**:742–757.

Astles, K. L., M. G. Holloway, A. Steffe, *et al.* (2006). An ecological method for qualitative risk assessment and its use in the management of fisheries in New South Wales, Australia. *Fisheries Research*, **82**:290–303.

Ayensu, E., D. van R. Claasen, M. Collins, *et al.* (1999). International ecosystem assessment. *Science*, **286**:685–686.

Baird, S. F. (1873). *Report on the Condition of the Sea Fisheries of the South Coast of New England in 1871 and 1872*. Washington, DC: Government Printing Office.

Beamish, R. J. (1993). Climate and exceptional fish production off the west coast of North America. *Canadian Journal of Fisheries and Aquatic Science*, **50**:2270–2291.

Beamish, R. J. and D. R. Bouillon (1993). Pacific salmon production trends in relation to climate. *Canadian Journal of Fisheries and Aquatic Science*, **50**:1002–1016.

Beamish, R. J., D. J. Noakes, G. A. McFarlane, *et al.* (1999). The regime concept and natural trends in the production of Pacific salmon. *Canadian Journal of Fisheries and Aquatic Science*, **56**:516–526.

Beddington, J. R. (1986). Shifts in resource populations in large marine ecosystems. In Sherman, K. and L. M. Alexander (eds.), *Variability and Management of Large Marine Ecosystems*. AAAS Selected Symposium 99. Washington, DC: AAAS.

Beddington, J. R. and G. P. Kirkwood (2005). The estimation of potential yield and stock status using life-history parameters. *Philosophical Transactions of the Royal Society B*, **360**:163–170.

Beddington, J. R., D. J. Agnew, and C. W. Clark (2007). Current problems in the management of marine fisheries. *Science*, **316**:1713–1716.

Begley, J. and D. Howell (2004). *An Overview of GADGET, Globally Applicable Area-Disaggregated General Ecosystem Toolbox*. ICES Document CM 2004/FF-13. Copenhagen: ICES.

Berkes, F. (1999). *Sacred Ecology: Traditional Ecological Knowledge and Management Systems*. London: Taylor & Francis.

Berkes, F. and C. Folke (1998). *Linking Social and Ecological Systems*. Cambridge, UK: Cambridge University Press.

Berkes, F., T.P. Hughes, R.S. Steneck, *et al.* (2006). Globalization, roving bandits, and marine resources. *Science*, **311**:1557–1558.

Beverton, R. (1998). Fish, fact and fantasy: a long view. *Reviews in Fish Biology and Fisheries*, **8**:229–249.

Beverton, R.J.H. and S.J. Holt (1957). *On the Dynamics of Exploited Fish Populations*. Fisheries Investment Series 2, Vol. 19. London: UK Ministry of Agriculture and Fisheries.

Bilio, M. and U. Niermann (2004). Is the comb jelly really to blame for it all? *Mnemiopsis leidyi* and the ecological concerns about the Caspian Sea. *Marine Ecology Progress Series*, **269**:173–183.

Bogstad, B., K. H. Hauge, and Ø. Ulltang (1997). MULTSPEC: a multi-species model for fish and marine mammals in the Barents Sea. *Journal of Northwest Atlantic Fishery Science*, **22**:317–341.

Boldt, J. (ed.) (2008). *Ecosystem Considerations for 2009*. Appendix C. Anchorage, AK: North Pacific Fishery Management Council.

Borley, J.O., E.S. Russell, M. Graham, W. Wallace, and D.E. Thursby-Pellam (1923). *The Plaice Fishery and the War: Preliminary Report on Investigations*. Fisheries Investigations (Series 2) 5. London: UK Ministry of Agriculture, Fisheries and Food.

Botsford, L.W., J.C. Castilla, and C.H. Peterson (1997). The management of fisheries and marine ecosystems. *Science*, **277**:509–515.

Box, G. and N. Draper (1987). *Empirical Model-Building and Response Surfaces*. Wiley.

Boyd, J. and S. Banzhaf (2007). What are ecosystem services? The need for standardized environmental accounting units. *Ecological Economics*, **63**:616–626.

Braccini, J.M., B.M. Gillanders, and T.I. Walker (2006). Hierarchical approach to the assessment of fishing effects on non-target chondricthyans: case study of *Squalus megalops* in southeastern Australia. *Canadian Journal of Fisheries and Aquatic Science*, **63**:2456–2466.

Brand, E.J., I.C. Kaplan, C.J. Harvey, *et al.* (2007). *A Spatially Explicit Ecosystem Model of the California Current's Food Web and Oceanography*. NOAA Technical Memorandum NMFS-NWFSC-84. 145 pp.

Brander, K.M. (1994). Patterns of distribution, spawning, and growth in North Atlantic cod: the utility of inter-regional comparisons. *ICES Marine Science Symposium*, **198**:89–103.

Brander, K.M. (1995). The effect of temperature on growth of Atlantic cod (*Gadus morhua* L.). *ICES Journal of Marine Science*, **52**:1–10.

Brander, K.M. (2007). Global fish production and climate change. *Proceedings of the National Academy of Sciences*, **104**:19 709–19 714.

Brander, K.M., G. Blom, M.F. Borges, *et al.* (2003). Changes in fish distribution in the eastern North Atlantic: are we seeing a coherent response to changing temperature? *ICES Marine Science Symposium*, **219**:261–270.

Brodeur, R.D., H. Sugisaki, and G.L. Hunt (2002). Increases in jellyfish biomass in the Bering Sea: implications for the ecosystem. *Marine Ecology Progress Series*, **233**:89–103.

Brodziak, J. and J. Link (2002). Ecosystem-based fishery management: what is it and how can we do it? *Bulletin of Marine Science*, **70**:589–611.

Browman, H.I. and K.I. Stergiou (2004). Perspectives on ecosystem-based approaches to the management of marine resources. *Marine Ecology Progress Series*, **274**:269–303.

Browman, H.I. and K.I. Stergiou (2005). Politics and socio-economics of ecosystem-based management of marine resources. *Marine Ecology Progress Series*, **300**:241–296.

Brown, B., J. Breenan, M. Grosslein, E. Heyerdahl, and R. Hennemuth (1976). The effect of fishing on the marine finfish biomass in the Northwest Atlantic from the Gulf of Maine to Cape Hatteras. *ICNAF Research Bulletin*, **12**:49–68.

Bundy, A. (2001). Fishing on ecosystems: the interplay of fishing and predation in Newfoundland-Labrador. *Canadian Journal of Fisheries and Aquatic Science*, **58**:1153–1167.

Bundy, A., R. Chuenpagdee, S. Jentoft, and R. Mahon (2008). If science is not the answer, what is? An alternative governance model for the world's fisheries. *Frontiers in Ecology and the Environment*, **6**:152–155.

Burnham, K.P. and D. Anderson (2002). *Model Selection and Multi-Model Inference: A Practical Information-Theoretic Approach*. New York: Springer-Verlag.

Burnside, D. and G.A. Rasmussen (1997). Ecosystem management: can it succeed? *Rangelands*, **19**(2):20–24.

Butterworth, D.S. (2007). Why a management procedure approach? Some positives and negatives. *ICES Journal of Marine Science*, **64**:613–617.

Butterworth, D.S. and A.E. Punt (1999). Experiences in the evaluation and implementation of management procedures. *ICES Journal of Marine Science*, **56**:985–998.

Butterworth, D.S., K.L. Cochrane, and J.A.A. de Oliveria (1997). Management procedures: a better way to manage fisheries? The South African experience. In Pikitch, E.K., D.D. Huppert, and M.P. Sissenwine (eds.) *Global Trends: Fisheries Management*, American Fisheries Society Symposium 20. Bethesda, MD: American Fisheries Society.

Caddy, J.F. (1999). Fisheries management in the twenty-first century: will new paradigms apply? *Reviews in Fish Biology and Fisheries*, **9**:1–43.

Caddy, J.F. (2002). Limit reference points, traffic lights, and holistic approaches to fisheries management with minimal stock assessment input. *Fisheries Research*, **56**:133–137.

Caddy, J.F. and D.J. Agnew (2004). An overview of recent global experience with recovery plans for depleted marine resources and suggested guidelines for recovery planning. *Reviews in Fish Biology and Fisheries*, **14**:43–112.

Caddy, J.F. and K.L. Cochrane (2001). A review of fisheries management past and present and some future perspectives for the third millennium. *Ocean and Coastal Management*, **44**:653–682.

Cadrin, S.X. (1999). A precautionary approach to fishery control rules based on surplus production modeling. In Restrepo, V.R. (ed.), *Proceedings of the Fifth National NMFS Stock Assessment Workshop: Providing Scientific Advice to Implement the Precautionary Approach Under the Magnuson-Stevens Fishery Conservation and Management Act*. NOAA Technical Memorandum NMFS-F/SPO 40, pp. 17–22.

Callicott, J.B. (1994). *Earth's Insights: A Multicultural Survey of Ecological Ethics From the Mediterranean Basin to the Australian Outback*. Berkeley, CA: University of California Press.

Campostrini, P. (2007). *Scientific Research and Safeguarding of Venice. Research Programme 2004–2006*. Vol. V: *2005 Results*. Venice, Italy: CORILA.

Canter, L.W. (2001). *Environmental Impact Assessment*. New York: McGraw-Hill.

Carson, R. (1962). *Silent Spring*. Boston, MA: Houghton Mifflin.

CBFEAP (Chesapeake Bay Fishery Ecosystem Advisory Panel) (2006). *Fisheries Ecosystem Planning for Chesapeake Bay*. Vol. 3. Bethesda, MD: American Fisheries Society.

Charles, A. (2001). *Sustainable Fishery Systems*. Malden, MA: Blackwell Science.

Chavez, F. P., J. Ryan, S. E. Lluch-Cota, and M. Niquen (2003). From anchovies to sardines and back: multidecadal change in the Pacific Ocean. *Science*, **299**:217–221.

Cheung, W. W. L., T. J. Pitcher, and D. Pauly (2005). A fuzzy logic expert system to estimate intrinsic extinction vulnerabilities of marine fishes to fishing. *Biological Conservation*, **124**:97–111.

Christensen, N. L., A. M. Bartuska, J. H. Brown, *et al.* (1996). The report of the Ecological Society of America Committee on the scientific basis for ecosystem management. *Ecological Applications*, **6**:665–691.

Christensen, V. and D. Pauly (1992). On steady-state modeling of ecosystems. In Christensen, V. and D. Pauly (eds.), *Trophic Models of Aquatic Ecosystems: ICLARM Conference Proceedings 26*. Manila, Philippines: ICLARM, pp. 14–19.

Christensen, V. and D. Pauly (2004). Placing fisheries in their ecosystem context, an introduction. *Ecological Modelling*, **172**:103–107.

Christy, F. T. (1996). The death rattle of open access and the advent of property rights in fisheries. *Marine Resource Economics*, **11**:287–304.

Clarke, G. L. (1946). Dynamics of production in a marine area. *Ecological Monographs*, **16**:323–325.

Colding, J. and C. Folke (2001). Social taboos: "invisible" systems of local resource management and biological conservation. *Ecological Applications*, **11**:584–600.

Coll, M., S. Libralato, S. Tudela, I. Palomera, and F. Pranovi (2008). Ecosystem overfishing in the ocean. *Public Library of Science (PLOS) ONE*, **3**(12): e3881. doi:10.1371/journal.pone.0003881.

Coll, M., L. J. Shannon, D. Yemane, *et al.* (2010). Ranking ecological relative status of exploited ecosystems using multiple indicators. *ICES Journal of Marine Science*, **67**: 769–786.

Collie, J. S. and A. K. DeLong (1999). Multispecies interactions in the Georges Bank fish community. In *Ecosystem Approaches for Fisheries Management*. Vol. AK-SG-99-01. Fairbanks, AK: Alaska Sea Grant College Program, pp. 187–210.

Collie, J. S. and H. Gislason (2001). Biological reference points for fish stocks in a multispecies context. *Canadian Journal of Fisheries and Aquatic Science*, **58**:2167–2176.

Collie, J. S., G. A. Escanero, and P. C. Valentine (1997). Effects of bottom fishing on the benthic megafauna of Georges Bank. *Marine Ecology Progress Series*, **155**:159–172.

Collie, J. S., S. J. Hall, M. J. Kaiser, and I. R. Poiner (2000). A quantitative analysis of fishing impacts on shelf-sea benthos. *Journal of Animal Ecology*, **69**:785–798.

Collie, J. S., H. Gislason, and M. Vinther (2003). Using AMOEBAs to display multispecies, multifleet fisheries advice. *ICES Journal of Marine Science*, **60**:709–720.

Conover, D. O. and S. B. Munch (2002). Sustaining fisheries yields over evolutionary time scales. *Science*, **297**:94–96.

Constable, A. J. (2001). The ecosystem approach to managing fisheries: achieving conservation objectives for predators of fished species. *CCAMLR Science*, **8**:37–64.

Constable, A. J., W. K. de la Mare, D. J. Agnew, I. Everson, and D. Miller (2000). Managing fisheries to conserve the Antarctic marine ecosystem: practical implementation of the Convention for the Conservation of Antarctic Marine Living Resources (CCAMLR). *ICES Journal of Marine Science*, **57**:778–791.

Costanza, R., R. d'Arge, R. de Groot, *et al.* (1997). The value of the world's ecosystem services and natural capital. *Nature*, **387**:253–260.

Costanza, R., F. Andrade, P. Antunes, *et al.* (1998). Principles for sustainable governance of the oceans. *Science*, **281**:198–199.

Crowder, L.B., G. Osherenko, O.R. Young, *et al.* (2006). Resolving mismatches in U.S. Ocean Governance. *Science*, **313**:617–618.

Cury, P.M. (2004). Tuning the ecoscope for the ecosystem approach to fisheries. In Browman, H.-I. and K.-I. Stergiou (eds.), Perspectives on ecosystem-based approaches to the management of marine resources. *Marine Ecology Progress Series*, **274**: 272–275.

Daan, N. and M.-P. Sissenwine eds. (1991). *Multispecies Models Relevant to Management of Living Resources*. Proceedings of ICES Marine Science Symposium 193. Copenhagen: ICES.

Dale, V.H., S. Brown, R.A. Haeuber, *et al.* (2000). Ecological principles and guidelines for managing the use of land. *Ecological Applications*, **10**:639–670.

Dambacher, J.M., H.W. Li, and P.A. Rossignol (2002). Relevance of community structure in assessing indeterminacy of ecological predictions. *Ecology*, **83**:1372–1385.

Dame, J.K. and R.R. Christian (2006). Uncertainty and the use of network analysis for ecosystem-based fishery management. *Fisheries*, **31**:331–341.

Davis, A.J., J.M. Hanson, H. Watts, and H. MacPherson (2004). Local ecological knowledge and marine fisheries research: the case of white hake (*Urophycis tenuis*) predation on juvenile American lobster (*Homarus americanus*). *Canadian Journal of Fisheries and Aquatic Science*, **61**:1191–1201.

Dayton, P.K. (1998). Reversal of the burden of proof in fisheries management. *Science*, **279**:821–822.

de la Mare, W.K. (1986). Simulation studies on management procedures. *Reports of the International Whaling Commission*, **36**:429–450.

de la Mare, W.K. (1998). Tidier fisheries management requires a new MOP (management-oriented paradigm). *Reviews in Fish Biology and Fisheries*, **8**:349–356.

de la Mare, W.K. (2005). Marine ecosystem-based management as a hierarchical control system. *Marine Policy*, **29**:57–68.

Deroba, J.J. and J.R. Bence (2008). A review of harvest policies: understanding relative performance of control rules. *Fisheries Research*, **94**:210–223.

Diamond, J. (2005). *Guns, Germs and Steel: The Fates of Human Societies*. New York: W.W. Norton.

Dichmont, C.M., A.R. Deng, A.E. Punt, W. Venables, and M. Haddon (2006a). Mangement strategies for short-lived species: the case of Australia's Northern Prawn Fishery. 1. Accounting for multiple species, spatial structure and implementation uncertainty when evaluating risk. *Fisheries Research*, **82**:204–220.

Dichmont, C.M., A.R. Deng, A.E. Punt, W. Venables, and M. Haddon (2006b). Management strategies for short lived species: the case of Australia's northern prawn fishery. 2. Choosing appropriate management strategies using input controls. *Fisheries Research*, **82**:221–234.

DOC (Department of Commerce) (2005). *Fisheries of the United States, 2004*. Silver Spring, MD: U.S. Department of Commerce, National Oceanic and Atmospheric Administration, National Marine Fisheries Service, Office of Science and Technology.

DOC (Department of Commerce) (2008). *Fisheries of the United States, 2007*. Silver Spring, MD: U.S. Department of Commerce, National Oceanic and Atmospheric Administration, National Marine Fisheries Service, Office of Science and Technology.

Doubleday, W.G. and D. Rivard (eds.) (1981) *Bottom Trawl Surveys*. Canadian Special Publication of Fisheries and Aquatic Sciences, 58.

Drinkwater, K.F. (2002). A review of the role of climate variability in the decline of northern cod. *American Fisheries Society Symposium*, **32**:113–130.

Drinkwater, K. F. (2005). The response of Atlantic cod (*Gadus morhua*) to future climate change. *ICES Journal of Marine Science*, **62**:1327–1337.
Dulvy, N. K. and J. D. Reynolds (2002). Predicting extinction vulnerability in skates. *Conservation Biology*, **16**:440–450.
Dulvy, N. K., Y. Sadovy, and J. D. Reynolds (2003). Extinction vulnerability in marine populations. *Fish and Fisheries*, **4**:25–64.
EcoAP (Ecosystem Assessment Program) (2009). *Ecosystem Status Report for the Northeast U.S. Continental Shelf Large Marine Ecosystem*. Northeast Fisheries Science Center Reference Document 09–11, 34 pp.
Edwards, S. F. (2003). Property rights to multi-attribute fishery resources. *Ecological Economics*, **44**:309–323.
Edwards, S. F., J. S. Link, and B. P. Rountree(2004). Portfolio management of wild fish stocks. *Ecological Economics*, **49**:317–329.
EFEP (European Fisheries Ecosystem Plan) (2004). *The North Sea Fisheries Ecosystem Plan, EU Quality of Life and Management of Living Resources*. QLRT-CT-2001-01685. Newcastle upon Tyne, UK: University of Newcastle.
Ellison, A. M. (1996). An introduction to Bayesian inference for ecological research and environmental decision-making. *Ecological Applications*, **6**:1036–1046.
Elton, C. and M. Nicholson (1942). The ten year cycle in numbers of the lynx in Canada. *Journal of Animal Ecology*, **11**:215–244.
Endter-Wada, J., D. Blahna, R. Krannich, and M. Brunson (1998). A framework for understanding social science contributions to ecosystem management. *Ecological Applications*, **8**:891–904.
Essington, T. E., A. H. Beaudreau, and J. Wiedenmann (2006). Fishing through marine food webs. *Proceedings of the National Academy of Sciences*, **103**:3171–3175.
FAO (Food and Agriculture Organization) (1995). *Precautionary Approach to Fisheries. Part 1: Guidelines on the Precautionary Approach to Capture Fisheries and Species Introductions*. FAO Fisheries Technical Paper 350, Part 1. Rome, Italy: FAO, 52 pp.
FAO (Food and Agriculture Organization) (2003). *The Ecosystem Approach to Fisheries*. FAO Technical Guidelines for Responsible Fisheries 4, Suppl. 2. Rome, Italy: FAO, 112 pp.
FAO (Food and Agriculture Organization) (2007). *The State of the World Fisheries and Aquaculture 2006*. Rome, Italy: FAO, 162 pp.
Fennel, W. and T. Neumann (2004). *Introduction to the Modeling of Marine Ecosystems*. Amsterdam, The Netherlands: Elsevier, 72 pp.
Ferrero, R. C. and L. W. Fritz (1994). *Comparisons of Walleye Pollock,* Theragra chalcogramma, *Harvest to Steller Sea Lion,* Eumetopias jubatus, *Abundance in the Bering Sea and Gulf of Alaska*. NOAA Technical Memorandum NMFS-AFSC-43.
Field, J. C. and R. C. Francis (2006). Considering ecosystem-based fisheries management in the California Current. *Marine Policy*, **30**:552–569.
Fletcher, W. J. (2005). The application of qualitative risk assessment methodology to prioritize issues for fisheries management. *ICES Journal of Marine Science*, **62**:1576–1587.
Fletcher, W. J., J. Chesson, K. J. Sainsbury, T. J. Hundloe, and M. Fisher (2005). A flexible and practical framework for reporting on ecologically sustainable development for wild capture fisheries. *Fisheries Research*, **71**:175–183.
Fogarty, M. J. (2005). Impacts of fishing activities on benthic habitat and carrying capacity: approaches to assessing and managing risk. *American Fisheries Society Symposium*, **41**:769–784.
Fox, W. W. (1975). Fitting the generalized stock production model by least-squares and equilibrium approximation. *Fishery Bulletin*, **73**:23–37.

Francis, R.C., S.R. Hare, A.B. Hollowed, and W.S. Wooster (1998). Effects of interdecadal climate variability on the oceanic ecosystems of the NE Pacific. *Fisheries Oceanography* **7**:1–21.

Francis, R.C., M.A. Hixon, M.E. Clarke, S.A. Murawski, and S. Ralston (2007). Ten commandments for ecosystem-based fisheries scientists. *Fisheries*, **32**:217–233.

Francis, R.I.C.C. (1992). Use of risk analysis to assess fishery management strategies: a case study using orange roughy (*Hoplostethus atlanticus*) on the Chatham Rise, New Zealand. *Canadian Journal of Fisheries and Aquatic Science*, **49**:922–930.

Francis, R.I.C.C. and R. Shotton (1997). "Risk" in fisheries management: a review. *Canadian Journal of Fisheries and Aquatic Science*, **54**:1699–1715.

Franklin, J.F. (1997). Ecosystem management: an overview. In Boyce, M.S. and A. Haney (eds.), *Ecosystem Management: Applications for Sustainable Forest and Wildlife Resources*. New Haven, CT: Yale University Press, pp. 21–53.

Franz, E.H. (2001). Ecology, values, and policy. *BioScience*, **51**:469–474.

Frederick, S.W. and R.M. Peterman (1995). Choosing fisheries harvest policies: when does uncertainty matter? *Canadian Journal of Fisheries and Aquatic Science*, **52**:291–306.

Fulton, E.A., A.D.M. Smith, and C.R. Johnson (2003). Effect of complexity on marine ecosystem models. *Marine Ecology Progress Series*, **253**:1–16.

Fulton, E.A., J.S. Parslow, A.D.M. Smith, and C.R. Johnson (2004a). Biogeochemical marine ecosystem models. 2. The effect of physiological detail on model performance. *Ecological Modelling*, **173**:371–406.

Fulton, E.A., M. Fuller, A.D.M. Smith, and A.E. Punt (2004b). *Ecological Indicators of the Ecosystem Effects of Fishing: Final Report*. Report, R99/1546. Canberra, Australia: Australian Fisheries Management Authority.

Fulton, E.A., A.D.M. Smith, and A.E. Punt (2005). Which ecological indicators can robustly detect effects of fishing? *ICES Journal of Marine Science*, **62**:540–551.

Fulton, E.A., T. Morato, and T.J. Pitcher (2007). Modelling seamount ecosystems and their fisheries. In Pitcher, T.J., T. Morato, P.J.B. Hart, *et al.* (eds.), *Seamounts: Ecology, Fisheries and Conservation*. Fish and Aquatic Resources Series 12. Oxford, UK: Blackwell, pp. 296–332.

Furness, R.W. (1999). Does harvesting a million metric tons of sand lance per year from the North Sea threaten seabird populations? In *Ecosystem Approaches for Fisheries Management*. Vol. AK-SG-99-01. Fairbanks, AK: Alaska Sea Grant College Program, pp. 407–424.

Furness, R.W. and M.L. Tasker (2000). Seabird-fishery interactions: quantifying the sensitivity of seabirds to reductions in sand eel abundance, and identification of key areas for sensitive seabirds in the North Sea. *Marine Ecology Progress Series*, **202**:253–264.

Gabriel, W.L. and P.M. Mace (1999). A review of biological reference points in the context of the precautionary approach. In Restrepo, V.R. (ed.), *Proceedings of the Fifth National NMFS Stock Assessment Workshop: Providing Scientific Advice to Implement the Precautionary Approach Under the Magnuson-Stevens Fishery Conservation and Management Act*. NOAA Technical Memorandum NMFS-F/SPO 40, pp. 34–45.

Gamble, R.J. and J.S. Link (2009). Analyzing the tradeoffs among ecological and fishing effects on an example fish community: a multispecies (fisheries) production model. *Ecological Modelling*, **220**(19):2570–2582.

Garcia, S.M. and K.L. Cochrane (2005). Ecosystem approach to fisheries: a review of implementation guidelines. *ICES Journal of Marine Science*, **62**:311–318.

Garcia, S. M. and I. de Leiva Moreno (2003). Global overview of marine fisheries. In Sinclair, M. and G. Valdimarsson (eds.), *Responsible Fisheries in the Marine Ecosystem*. Wallingford, UK: CAB International.

Garcia, S. M., A. Zerbi, C. Aliaume, T. Do Chi, and G. Lasserre (2003). *The Ecosystem Approach to Fisheries: Issues, Terminology, Principles, Institutional Foundations, Implementation and Outlook*. FAO Fisheries Technical Paper 443. Rome, Italy: FAO, 71 pp.

Garrison, L. P. and J. S. Link (2004). *An Expanded Multispecies Virtual Population Analysis (MSVPA-X) to Evaluate Predator-Prey Interactions in Exploited Fish Ecosystems*. User's Manual and Model Description, Version 1.1. Washington, DC: Atlantic States Marine Fisheries Commission.

Garrison, L. P., J. S. Link, D. P. Kilduff, *et al.* (2010). An expansion of the MSVPA approach for quantifying predator-prey interactions in exploited fish communities. *ICES Journal of Marine Science*, **67**, doi:10.1093/icesjms/fsq005.

GBRMPA (Great Barrier Reef Marine Park Authority) (2008). *Annual Report, 2007–2008*. Townsville, Queensland, Australia.

Gislason, H. and T. Helgason (1985). Species interactions in assessment of fish stocks with special application to the North Sea. *Dana*, **5**:1–44.

Gislason, H., M. Sinclair, K. Sainsbury, and R. O'Boyle (2000). Symposium overview: incorporating ecosystem objectives within fisheries management. *ICES Journal of Marine Science*, **57**:468–475.

GOOS (Global Ocean Observing System). (2007). Available www.ioc-goos.org. Accessed December 2, 2007.

Grafton, R. Q., R. Arnason, T. Bjørndal, *et al.* (2006). Incentive-based approaches to sustainable fisheries. *Canadian Journal of Fisheries and Aquatic Science*, **63**:699–710.

Graham, M. (1943). *The Fish Gate*. London: Faber.

Graham, M. (1948). Reporter's view of the scientific meeting on the effect of the war on the stocks of commercial food fishes. *Rapports, Conseil Permanent International pour l'Exploration de la Mer*, **132**:72–78.

Graham, M. (1956). Concepts of conservation. In *Papers Presented at the International Technical Conference on the Conservation of the Living Resources of the Sea*. United Nations Publication 1956.II.B.1, New York: United Nations, pp. 1–11.

Gray, T. S. (ed.) (2005). *Participation in Fisheries Governance*. Dordrecht, The Netherlands: Springer.

Griffiths, S. P., D. T. Brewer, D. S. Heales, D. A. Milton, and I. C. Stobutzki (2006). Validating ecological risk assessments for fisheries: assessing the impacts of turtle excluder devices on elasmobranch bycatch populations in an Australian trawl fishery. *Marine and Freshwater Research*, **57**:395–401.

Grumbine, R. E. (1994). What is ecosystem management? *Conservation Biology*, **8**:27–38.

Grumbine, R. E. (1997). Reflections on "What is ecosystem management?" *Conservation Biology*, **11**:41–47.

Gucu, A. C. (2002). Can overfishing be responsible for the successful establishment of *Mnemiopsis leidyi* in the Black Sea? *Estuarine, Coastal and Shelf Science*, **54**:439–451.

Gulland, J. A. (1983). *Fish Stock Assessment: A Manual of Basic Methods*. New York: Wiley.

Gunderson, L. H. and C. S. Holling, (eds.) (2002). *Panarchy: Understanding Transformations in Human and Natural Systems*. Washington, DC: Island Press.

Haeuber, R. (1996). Setting the environmental policy agenda: the case of ecosystem management. *Natural Resources Journal*, **36**:1–28.

Hall, S. J. (1999). *The Effects of Fishing on Marine Ecosystems and Communities*. Malden, MA: Blackwell Science.

Hall, S.J. and B. Mainprize (2004). Towards ecosystem-based fisheries management. *Fish and Fisheries*, **5**:1–20.

Halliday, R.G., P.F. Fanning, and R.K. Mohn (2001). *Use of the Traffic Light Method in Fishery Management Planning*. Canadian Science Advisory Secretariat Research Document 2001/108, 42 pp.

Hanna, S.S. (1998). Institutions for marine ecosystems: economic incentives and fishery management. *Ecological Applications*, **8**:S170–S174.

Hanna, S. (2001). Managing the human-ecological interface: marine resources as example and laboratory. *Ecosystems*, **4**:736–741.

Hanson, J.M. and M. Lanteigne (2000). Evaluation of Atlantic cod predation on American lobster in the southern Gulf of St. Lawrence, with comments on other potential fish predators. *Transactions of the American Fisheries Society*, **129**:13–29.

Hardin, G. (1968). The tragedy of the commons. *Science*, **162**:1243–1248.

Hardwood, J. (2000). Risk assessment and decision analysis in conservation. *Biological Conservation*, **95**:219–225.

Hare, J.A. and K.W. Able (2007). Mechanistic links between climate and fisheries along the east coast of the United States: explaining population outbursts of Atlantic croaker (*Micropogonias undulatus*). *Fisheries Oceanography*, **16**:31–45.

Hayes, D.B., C.P. Ferreri, and W.W. Taylor (1996). Linking fish habitat to their population dynamics. *Canadian Journal of Fisheries and Aquatic Science*, **53**: 383–390.

Helgason, T. and H. Gislason (1979). *VPA-analysis with Species Interaction Due to Predation*. ICES Document CM 1979/G:52, 193 pp.

Hennessey, T.M. and J.G. Sutinen (eds.) (2005). *Sustaining Large Marine Ecosystems: The Human Dimension*. Amsterdam: Elsevier Science.

Heymans, J.J. and D. Baird (2000). Network analysis of the northern Benguela ecosystem by means of NETWRK and ECOPATH. *Ecological Modelling*, **131**:97–119.

Hickling, C.F. (1946). The recovery of a deep sea fishery. In *UK Ministry of Agriculture and Fisheries, Fisheries Investigations* (Series 2), **17**:1–59.

Hilborn, R. (2004). Ecosystem-based fisheries management: the carrot or the stick? In Browman, H.I. and K.I. Stergiou (eds.), Perspectives on ecosystem-based approaches to the management of marine resources. *Marine Ecology Progress Series*, **274**:275–278.

Hilborn, R. (2006). Faith-based fisheries. *Fisheries*, **31**:554–555.

Hilborn, R. (2007). Managing fisheries is managing people: what has been learned? *Fish and Fisheries*, **8**:285–296.

Hilborn, R. and M. Mangel (1997). *The Ecological Detective: Confronting Models with Data*. Princeton, NJ: Princeton University Press.

Hilborn, R. and C.J. Walters (1992). *Quantitative Fisheries Stock Assessment: Choice, Dynamics and Uncertainty*. New York: Chapman and Hall.

Hjort, J. (1914). Fluctuations in the great fisheries of Northern Europe, viewed in the light of biological research. *Rapports et Procès-Verbaux des Réunions du Conseil Permanent International pour l'Exploration de la Mer*, **20**:1–228.

Hobday, A.J., A. Smith, I. Stobutzki, *et al.* (2004). Ecological risk assessment for Australian Commonwealth fisheries. Final Report Stage 1. Hazard identification and preliminary risk assessment. Report R01/0934. Hobart, Tasmania: CSIRO Marine Research, 74 pp.

Hobday, A.J., A. Smith, H. Webb, *et al.* (2007). Ecological risk assessment for the effects of fishing: methodology. Report R04/1072. for the Australian Fisheries Management Authority, Canberra, Australia, 186 pp.

Hoenig, J.M. (1983). Empirical use of longevity data to estimate mortality rates. *Fishery Bulletin*, **82**:898–902.

Holland, D. S. (2002). Integrating marine protected areas into models for fishery assessment and management. *Natural Resource Modeling*, **15**:369–386.

Holling, C. S. (1978). *Adaptive Environmental Assessment and Management*. New York: Wiley.

Holling, C. S. (2001). Understanding the complexity of economic, ecological, and social systems. *Ecosystems*, **4**:390–405.

Hollowed, A. B. and W. S. Wooster (1995). Decadal-scale variations in the eastern subarctic Pacific. 2. Response of northeast Pacific fish stocks. *Canadian Special Publication in Fisheries and Aquatic Sciences* **121**:373–385.

Hollowed, A. B., E. Brown, J. Ianelli, *et al.* (1997). *Walleye Pollock*. Stock Assessment and Fishery Evaluation Report for Gulf of Alaska as Projected for 1998. Anchorage, AK: North Pacific Fisheries Management Council.

Hollowed, A. B., N. Bax, R. Beamish, *et al.* (2000a). Are multispecies models an improvement on single-species models for measuring fishing impacts on marine ecosystems? *ICES Journal of Marine Science*, **57**:707–719.

Hollowed, A. B., J. N. Ianelli, and P. A. Livingston (2000b). Including predation mortality in stock assessments: a case study for Gulf of Alaska walleye pollock. *ICES Journal of Marine Science*, **57**:279–293.

Hollowed, A. B., S. R. Hare, and W. S. Wooster (2001). Pacific Basin climate variability and patterns of Northeast Pacific marine fish production. *Progress in Oceanography*, **49**:257–282.

Hutchings, J. A. (2000). Collapse and recovery of marine fishes. *Nature*, **406**:882–885.

Hutchings, J. A. and J. D. Reynolds (2004). Marine fish population collapses: consequences for recovery and extinction risk. *BioScience*, **54**:297–309.

Huxley, T. H. (1884). Inaugural address. *Fisheries Exhibition Literature*, **4**:1–22.

ICES (International Council for the Exploration of the Sea) (2003). *Report of the Working Group on Ecosystem Effects of Fishing Activities*. ICES Document CM 2003/ACE:05, 193 pp.

ICES (International Council for the Exploration of the Sea) (2008). *Report of the Working Group on Multispecies Assessment Methods (WGSAM), 6–10 October 2008, ICES Headquarters, Copenhagen*. ICES Document CM 2008/RMC:06.

IOOS (Integrated Ocean Observing System) (2007). Available www.ocean.us/ioos_observing_system. Accessed December 2, 2007.

IPCC (2007). *Climate Change 2007: The Physical Science Basis. Contribution of Working Group I to the Fourth Assessment Report of the Intergovernmental Panel on Climate Change*. Cambridge, UK: Cambridge University Press.

IWC (International Whaling Commission) (2008). *Report of the International Whaling Commission, Volume 60*. Cambridge, UK: IWC.

Jackson, J. B. C., *et al.* (2001). Historical overfishing and the recent collapse of coastal ecosystems. *Science*, **293**:629–638.

Jacobson, L. D. and A. D. McCall (1995). Stock-recruitment models for Pacific sardine (*Sardinops sagax*). *Canadian Journal of Fisheries and Aquatic Science*, **52**:566–577.

Jacobson, L. D., S. J. Bograd, R. H. Parrish, R. Mendelssohn, and F. B. Schwing (2005). An ecosystem-based hypothesis for climatic effects on surplus production in California sardine (*Sardinops sagax*) and environmentally dependent surplus production models. *Canadian Journal of Fisheries and Aquatic Science*, **62**:1782–1796.

Jamieson, G., R. O'Boyle, J. Arbour, *et al.* (2001). *Proceedings of the National Workshop on Objectives and Indicators for Ecosystem-based Management*. Canadian Stock Assessment Secretariat (CSAS) Proceedings Series 2001/9, Ottawa, Canada: DFO, CSAS, 142 pp.

Jennings, S. (2005). Indicators to support an ecosystem approach to fisheries. *Fish and Fisheries*, **6**:212–232.

Jennings, S. and M.J. Kaiser (1998). The effects of fishing on marine ecosystems. *Advances in Marine Biology*, **34**:201–352.

Jennings, S., J.D. Reynolds, and N.V.C. Polunin (1999). Predicting the vulnerability of tropical reef fishes to exploitation with phylogenies and life histories. *Conservation Biology*, **13**:1466–1475.

Jorgensen, S.E. (1992). Exergy and ecology. *Ecological Modelling*, **63**:185–214.

Jurado-Molina, J. and P. Livingston (2002). Multispecies perspectives on the Bering Sea groundfish fisheries management regime. *North American Journal of Fisheries Management*, **22**:1164–1175.

Jurado-Molina, J., P. Livingston, and J. Ianelli (2005). Incorporating predation interactions in a statistical catch-at-age model for a predator-prey system in the Eastern Bering Sea. *Canadian Journal of Fisheries and Aquatic Science*, **62**:1865–1873.

Kaiser, M.J. (2003). Detecting the effects of fishing on seabed community diversity: importance of scale and sample size. *Conservation Biology*, **17**:512–520.

Kaiser, M.J. and S.J. de Groot (2000). *Effects of Fishing on Non-Target Species and Habitats*. Oxford, UK: Blackwell Science.

Kaiser, M.J. and G. Edwards-Jones (2006). The role of ecolabeling in fisheries management and conservation. *Conservation Biology*, **20**:392–398.

Kaiser, M.J., J.S. Collie, S.J. Hall, S. Jennings, and I.R. Poiner (2002). Modification of marine habitats by trawling activities: prognosis and solutions. *Fish and Fisheries*, **3**:114–136.

Kavanagh, P., N. Newlands, V. Christensen, and D. Pauly (2004). Automated parameter optimization for EcoPath ecosystem models. *Ecological Modelling*, **172**:141–149.

Keyl, F. and M. Wolff (2008). Environmental variability and fisheries: what can models do? *Reviews in Fish Biology and Fisheries*, **18**:273–299.

Kurtz, J.C., L.E. Jeckson, and W.S. Fisher (2001). Strategies for evaluating indicators based on guidelines from the Environmental Protection Agency's Office of Research and Development. *Ecological Indicators*, **1**:49–60.

Lackey, R.T. (1998). Seven pillars of ecosystem management. *Landscape and Urban Planning*, **40**:21–30.

Laevastu, T. and H.A. Larkin (1981). *Marine Fisheries Ecosystem: Its Quantitative Evaluation and Management*. Farnham, Surrey, UK: Fishing News Books.

Lane, D.E. and R.L. Stephenson (1998). A framework for risk analysis in fisheries decision-making. *ICES Journal of Marine Science*, **55**:1–13.

Larkin, P.A. (1977). An epitaph for the concept of maximum sustainable yield. *Transactions of the American Fisheries Society*, **106**:1–11.

Larkin, P.A. (1996). Concepts and issues in marine ecosystem management. *Reviews in Fish Biology and Fisheries*, **6**:139–164.

Leopold, A. (1949). *A Sand County Almanac*. Oxford, UK: Oxford University Press.

Leslie, H., A.A. Rosenberg, and J. Eagle (2008). Is a new mandate needed for marine ecosystem-based management? *Frontiers in Ecology and the Environment*, **6**:43–48.

Leslie, P.H. and J.C. Gower (1960). Properties of a stochastic model for the predator-prey type of interaction between two species. *Biometrika*, **47**:219–301.

Levin, P.S., M.J. Fogarty, S.A. Murawski, and D. Fluharty (2009). Integrated ecosystem assessments: developing the scientific basis for ecosystem-based management of the ocean. *Public Library of Science (PLOS) Biology*, **7**(1): e1000014 (0023–0028).

Libralato, S., V. Christensen, and D. Pauly (2006). A method for identifying keystone species in food web models. *Ecological Modelling*, **195**:153–171.

Libralato, S., M. Coll, S. Tudela, I. Palomera, and F. Pranovi (2008). Novel index for quantification of ecosystem effects of fishing as removal of secondary production. *Marine Ecology Progress Series*, **355**:107–129.

Lindeman, R. L. (1942). The trophic-dynamic aspect of ecology. *Ecology*, **23**:399–418.

Link, J. S. (2001). A framework for ecosystem-based fisheries management. In *7th National Stock Assessment Workshop (Re)Building Sustainable Fisheries and Marine Ecosystems*. Santa Cruz, CA: NMFS, SWFSC, 15 pp.

Link, J. S. (2002a). Does food web theory work for marine ecosystems? *Marine Ecology Progress Series*, **230**:1–9.

Link, J. S. (2002b). Ecological considerations in fisheries management: when does it matter? *Fisheries*, **27**:10–17.

Link, J. S. (2002c). What does ecosystem-based fisheries management mean? *Fisheries*, **27**:18–21.

Link, J. S. (2005). Translation of ecosystem indicators into decision criteria. *ICES Journal of Marine Science*, **62**:569–576.

Link, J. S. (2007). Under-appreciated species in ecology: "ugly fish" in the Northwest Atlantic Ocean. *Ecological Applications*, **17**:2037–2060.

Link, J. S. and F. P. Almeida (2000). *An Overview in History of the Food Web Dynamics Program of the Northeast Fisheries Science Center, Woods Hole, Massachusetts*. NOAA Technical Memorandum NMFS-NE-159, 60 pp.

Link, J. S. and J. Brodziak (eds.) (2002). *Status of the Northeast U.S. Continental Shelf Ecosystem: A Report of the Northeast Fisheries Science Center's Ecosystem Status Working Group*. Northeast Fisheries Science Center Reference Document 02–11, 245 pp.

Link, J. S. and J. S. Idoine (2009). Estimates of predator consumption of the northern shrimp *Pandalus borealis*, with implications for estimates of population biomass in the Gulf of Maine. *North American Journal of Fisheries Management*, **29**:1567–1583.

Link, J. S. and K. Sosebee (2008). Estimates and implications of skate consumption in the northeastern U.S. continental shelf ecosystem. *North American Journal of Fisheries Management*, **28**:649–662.

Link, J. S., J. K. T. Brodziak, S. F. Edwards, *et al.* (2002a). Marine ecosystem assessment in a fisheries management context. *Canadian Journal of Fisheries and Aquatic Science*, **59**:1429–1440.

Link, J. S., L. P. Garrison, and F. P. Almeida (2002b). Interactions between elasmobranchs and groundfish species (Gadidae and Pleuronectidae) on the Northeast U.S. shelf. I. Evaluating predation. *North American Journal of Fisheries Management*, **22**:550–562.

Link, J. S., F. Almeda, P. Valentine, *et al.* (2005a). The effect of area closures on Georges Bank. *AFS Symposium*, **41**:345–368.

Link, J. S., M. J. Fogarty, and R. W. Langton (2005b). The trophic ecology of flatfishes. In Gibson, R. N. (ed.), *Biology of Flatfishes*. Oxford, UK: Blackwell Science, pp. 185–212.

Link, J. S., W. T. Stockhausen, and E. T. Methratta (2005c). Food web theory in marine ecosystems. In Belgrano, A., U. M., Scharler, J. Dunne, and R. E. Ulanowicz (eds.), *Aquatic Food Webs: An Ecosystem Approach*. Oxford, UK: Oxford University Press.

Link, J. S., W. Overholtz, J. O'Reilly, *et al.* (2008). An overview of EMAX: the Northeast U.S. continental shelf ecological network. *Journal of Marine Systems*, **74**:453–474.

Link, J. S., L. Col, V. Guida, *et al.* (2009a). Response of balanced network models to large-scale perturbation: implications for evaluating the role of small pelagics in the Gulf of Maine. *Ecological Modelling*, **220**:351–369.

Link, J. S., B. Bogstad, H. Sparholt, and G. R. Lilly (2009b). The role of cod in the ecosystem. *Fish and Fisheries*, **10**:58–87.

Link, J.S., D. Yemane, L.J. Shannon, *et al.* (2010). Relating marine ecosystem indicators to fishing and environmental drivers: an elucidation of contrasting responses. *ICES Journal of Marine Science*, **67**:787–795.

Livingston, P. (ed.) (2002). *Ecosystem Considerations for 2003*. Groundfish Plan Team Document. Anchorage, AK: Management Council North Pacific Fishery.

Livingston, P. and J. Jurado-Molina (2000). A multispecies virtual population analysis of the eastern Bering Sea. *ICES Journal of Marine Science*, **57**:294–299.

Lotka, A.J. (1925). *Elements of Physical Biology*. Baltimore, MD: Williams and Wilkins.

Lotze, H.K. (2004). Repetitive history of resource depletion and mismanagement: the need for a shift in perspective. In Browman, H.I. and K.I. Stergiou (eds.) Perspectives on ecosystem-based approaches to the management of marine resources. *Marine Ecology Progress Series*, **274**: 282–285.

Lotze, H.K., H.S. Lenihan, B.J. Bourque, *et al.* (2006). Depletion, degradation, and recovery potential of estuaries and coastal seas. *Science*, **312**:1806–1809.

Loughlin, T.R. and R.L. Merrick (1989). Comparison of commercial harvest of walleye pollock and northern sea lion abundance in the Bering Sea and Gulf of Alaska. In *Proceedings of the International Symposium on the Biology and Management of Walleye Pollock, 14–16 November 1988, Anchorage, Alaska*. Vol. AK-SG-89–01. Fairbanks, AK: Alaska Sea Grant Program.

Ludwig, D. and C.J. Walters (1989). A robust method for parameter estimation from catch and effort data. *Canadian Journal of Fisheries and Aquatic Science*, **46**:137–144.

Ludwig, D., M. Mangel, and B. Haddad (2001). Ecology, conservation and public policy. *Annual Review of Ecology and Systematics*, **32**:481–517.

Mace, G.M., N.J. Collar, K.J. Gaston, *et al.* (2008). Quantification of extinction risk: IUCN's system for classifying threatened species. *Conservation Biology*, **22**:1424–1442.

Mace, P.M. (2001). A new role for MSY in single-species and ecosystem approaches to fisheries stock assessment and management. *Fish and Fisheries*, **2**:2–32.

Mace, P.M. (2004). In defense of fisheries scientists, single-species models and other scapegoats: confronting the real problems. In Browman, H.I. and K.I. Stergiou (eds.), Perspectives on ecosystem-based approaches to the management of marine resources. *Marine Ecology Progress Series*, **274**:285–291.

Magnusson, K.G. (1995). An overview of the multispecies VPA – theory and applications. *Reviews in Fish Biology and Fisheries*, **5**:195–212.

Malone, C.R. (2000). Ecosystem management policies in state government of the USA. *Landscape and Urban Planning*, **48**:57–64.

Mangel, M. and P.S. Levin (2005). Regime, phase and paradigm shifts: making community ecology the basic science for fisheries. *Philosophical Transactions of the Royal Society B*, **360**:95–105.

Mangel, M., L.M. Talbot, G.K. Meffe, *et al.* (1996). Principles for the conservation of wild living resources. *Ecological Applications*, **6**:338–362.

Marasco, R.J., D. Goodman, C.B. Grimes, *et al.* (2007). Ecosystem-based fisheries management: some practical suggestions. *Canadian Journal of Fisheries and Aquatic Science*, **64**:928–939.

May, R.M., J.R. Beddington, C.W. Clark, S.J. Holt, and R.M. Laws (1979). Management of multispecies fisheries. *Science*, **205**:267–277.

McCall, A.D. (1996). *Patterns of Low-frequency Variability in Fish Populations of the California Current. CalCOFI (California Cooperative Oceanic Fishery Investigations) Reports*, **37**:100–110.

Merrick, R.L., M.K. Chumbley, and G.V. Byrd (1997). Diet diversity of Steller sea lions (*Eumetopias jubatus*) and their population decline in Alaska: a potential relationship. *Canadian Journal of Fisheries and Aquatic Science*, **54**:1342–1348.

Methratta, E.T. and J.S. Link (2006). Evaluation of quantitative indicators for marine fish communities. *Ecological Indicators*, **6**:575–588.

Milton, D.A. (2001). Assessing the susceptibility to fishing of populations of rare trawl bycatch: sea snakes caught by Australia's Northern Prawn Fishery. *Biological Conservation*, **101**:281–290.

Moustahfid, H., J.S. Link, W.J. Overholtz, and M.C. Tyrell (2009). The advantage of explicitly incorporating predation mortality into age-structured stock assessment models: an application for Northwest Atlantic mackerel. *ICES Journal of Marine Science*, **66**:445–454.

Mueter, F.J. and M.A. Litzow (2008). Sea ice retreat alters the biogeography of the Bering Sea continental shelf. *Ecological Applications*, **18**:309–320.

Mueter, F.J. and B.A. Megrey (2006). Using multi-species surplus production models to estimate ecosystem-level maximum sustainable yields. *Fisheries Research*, **81**:189–201.

Mullon, C., P. Fréon, and P. Cury (2005). The dynamics of collapse in world fisheries. *Fish and Fisheries*, **6**:111–120.

Murawski, S.A. (1984). Mixed-species yield per recruitment analyses accounting for technological interactions. *Canadian Journal of Fisheries and Aquatic Science*, **41**:897–916.

Murawski, S.A. (2000). Definitions of overfishing from an ecosystem perspective. *ICES Journal of Marine Science*, **57**:649–658.

Murawski, S.A. (2007). Ten myths concerning ecosystem approaches to marine resource management. *Marine Policy*, **31**:681–690.

Murawski, S.A., R. Brown, H.-L. Lai, P.J. Rago, and L. Hendrickson (2000). Large-scale closed areas as a fishery-management tool in temperate marine systems: the Georges Bank experience. *Bulletin of Marine Science*, **66**:775–798.

Murphy, B.R. and W. Willis (eds.) (1996). *Fisheries Techniques*. Bethesda, MD: American Fisheries Society.

Musick, J.A. (1999). Criteria to define extinction risk in marine fishes. *Fisheries*, **24**:6–14.

Myers, R.A. and B. Worm (2003). Rapid worldwide depletion of predatory fish communities. *Nature*, **423**:280–283.

Myers, R.A., J.A. Hutchings, and N.J. Barrowman (1997). Why do fish stocks collapse? The example of cod in Atlantic Canada. *Ecological Applications*, **7**:91–106.

Myers, R.A., J.K. Baum, T.D. Shepherd, S.P. Powers, and C.H. Peterson (2007). Cascading effects of the loss of apex predatory sharks from a coastal ocean. *Science*, **315**:1846–1850.

NEFSC (Northeast Fisheries Science Center) (2006). *42nd Northeast Regional Stock Assessment Workshop (42nd SAW) Stock Assessment Report*. Part B: *Expanded Multispecies Virtual Population Analysis (MSVPA-X) Stock Assessment Model*. Northeast Fisheries Science Center Reference Document 06–09. Woods Hole, MA: NEFSC.

NEFSC (Northeast Fisheries Science Center) (2008). Assessment of 19 Northeast Groundfish Stocks through 2007. In *Report of the 3rd Groundfish Assessment Review Meeting (GARM III), Northeast Fisheries Science Center, Woods Hole, Massachusetts, August 4–8, 2008. Section 2.1*. Northeast Fisheries Science Center Reference Document 08–1, pp. 855–865.

Nelson, G.A., B.C. Chase, and J. Stockwell (2003). Food habits of striped bass (*Morone saxatilis*) in coastal waters of Massachusetts. *Journal of Northwest Atlantic Fishery Science*, **32**:1–25.

Nicholson, M. D. and S. Jennings (2004). Testing candidate indicators to support ecosystem-based management: the power of monitoring surveys to detect temporal trends in fish community metrics. *ICES Journal of Marine Science*, **61**:35–42.

NMFS (National Marine Fisheries Service) (1999). *Ecosystem-based Fishery Management.* A report to Congress by the Ecosystems Principles Advisory Panel. Silver Spring, MD: U.S. Department of Commerce.

NOAA CBO (Chesapeake Bay Fisheries Ecosystem Advisory Panel) (2006). *Fisheries Ecosystem Planning for Chesapeake Bay*. Bethesda, MD: American Fisheries Society.

Norton, B. G. and D. Noonan (2007). Ecology and valuation: big changes needed. *Ecological Economics*, **63**:664–675.

NPFMC (North Pacific Fishery Management Council) (2008). *Aleutian Islands Fishery Ecosystem Plan*. Anchorage, AK: NPFMC.

NRC (National Research Council) (1999). *Sustaining Marine Fisheries*. Washington, DC: National Academy Press.

O'Boyle, R. and G. Jamieson (2006). Observations on the implementation of ecosystem-based management: experiences on Canada's east and west coasts. *Fisheries Research*, **79**:1–12.

O'Boyle, R., M. Sinclair, P. Keizer, *et al.* (2005). Indicators for ecosystem-based management on the Scotian Shelf: bridging the gap between theory and practice. *ICES Journal of Marine Science*, **62**:598–605.

Odum, E. P. (1969). The strategy of ecosystem development. *Science*, **164**:262–270.

Odum, H. T. (1957). Trophic structure and productivity of Silver Springs, Florida. *Ecological Monographs*, **27**:55–112.

Orensanz, J. M. and G. S. Jamieson (1998). The assessment and management of spatially explicit stocks: an overview of the North Pacific Symposium on Invertebrate Stock assessment and management. In Jamieson, G. S. and A. Campbell (eds.), *Proceedings of the North Pacific Symposium on Invertebrate Stock Assessment and Management*. Canadian Special Publication in Fisheries and Aquatic Sciences, **125**: 441–459.

Ostrom, E., J. Burger, C. B. Field, R. B. Norgaard, and D. Policansky (1999). Revisiting the commons: local lessons, global challenges. *Science*, **284**:278–282.

Overholtz, W. J., L. D. Jacobson, and J. S. Link (2008). Developing an ecosystem approach for assessment advice and biological reference points for the Gulf of Maine–Georges Bank herring complex: adding the impact of predation mortality. *North American Journal of Fisheries Management*, **28**:247–257.

Pascoe, S. (2006). Food for thought. *ICES Journal of Marine Science*, **63**:1–3.

Patrick, W. S. and K. Damon-Randall (2008). Using a five-factored structured decision analysis to evaluate the extinction risk of Atlantic sturgeon (*Acipenser oxyrinchus oxyrinchus*). *Biological Conservation*, **141**:2906–2911.

Patrick, W. S., P. Spencer, O. Ormseth, *et al.* (2009). *Use of Productivity and Susceptibility Indices to Determine the Vulnerability of a Stock: With Example Applications to Six U.S. Fisheries*. Final Report of the National Marine Fisheries Service Vulnerability Evaluation Work Group. Silver Spring, MD.

Pauly, D. (2003). Ecosystem impacts of the world's marine fisheries. *Global Change Newsletter*, **55**:21–23.

Pauly, D. and V. Christensen (1995). Primary production required to sustain global fisheries. *Nature*, **374**:255–257.

Pauly, D., V. Christensen, J. Dalsgaard, R. Froese, and F. C. Torres, Jr. (1998). Fishing down marine food webs. *Science*, **279**:860–863.

Pauly, D., V. Christensen, S. Guénette, *et al.* (2002). Towards sustainability in world fisheries. *Nature*, **418**:689–695.

Pauly, D., J. Alder, E. Bennett, *et al.* (2003). The future for fisheries. *Science*, **302**:1359–1361.

Pella, J.J. and P.K. Tomlinson (1969). A generalized stock-production model. *Bulletin of the Inter-American Tropical Tuna Commission*, **13**:419–496.

Perez, M.A. (1990). *Review of Marine Mammal Population and Prey Information for Bering Sea Ecosystem Studies*. NOAA Technical Memorandum NMFS F/NWC-186, 81 pp.

Perry, A.L., P.J. Low, J.R. Ellis, and J.D. Reynolds (2005). Climate change and distribution shifts in marine fishes. *Science*, **308**:1912–1915.

Peterman, R.M. (1990). Statistical power analysis can improve fisheries research and management. *Canadian Journal of Fisheries and Aquatic Science*, **47**:2–15.

Peterman, R.M. (2004). Possible solutions to some challenges facing fisheries scientists and managers. *ICES Journal of Marine Science*, **61**:1331–1343.

Peterman, R.M. and J.L. Anderson (1999). Decision analysis: a method for taking uncertainties into account in risk-based decision making. *Human and Ecological Risk Assessment*, **5**:231–244.

Peters, C.N. and D.R. Marmorek (2001). Application of decision analysis to evaluate recovery actions for threatened Snake River spring and summer chinook salmon. *Canadian Journal of Fisheries and Aquatic Science*, **58**:2431–2466.

Pikitch, E.K., C. Santora, E.A. Babcock, *et al.* (2004). Ecosystem-based fishery management. *Science*, **305**:346–347.

Pitcher, T.J. (2001). Fisheries managed to rebuild ecosystems? Reconstructing the past to salvage the future. *Ecological Applications*, **11**:601–617.

Pitcher, T.J., D. Kalikoski, K. Short, D. Varkey, and G. Pramod (2009). An evaluation of progress in implementing ecosystem-based management of fisheries in 33 countries. *Marine Policy*, **33**:223–232.

Plagányi, É.E. (2007). *Models for an Ecosystem Approach to Fisheries*. FAO Fisheries Technical Paper 477. Rome, Italy: FAO, 108 pp.

POC (Pew Oceans Commission) (2004). *America's Living Oceans: Charting a Course for Sea Change: A Report to the Nation*. Arlington, VA: Pew Oceans Commission, 143 pp.

Polovina, J.J. (1984). Model of a coral reef ecosystem. I. The ECOPATH model and its application to French Frigate Shoals. *Coral Reefs*, **3**:1–11.

Pope, J.G. (1979). *A Modified Cohort Analysis in which Constant Mortality is Replaced by Estimates of Predation Levels*. ICES Document CM 1979/H:16, 8 pp.

Pope, J.G., D.S. MacDonald, N. Daan, J.D. Reynolds, and S. Jennings (2000). Gauging the impact of fishing mortality on non-target species. *ICES Journal of Marine Science*, **57**:689–696.

Prager, M.H. (1994). A suite of extensions to a nonequilibrium surplus–production model. *Fishery Bulletin*, **92**:374–389.

Punt, A. and D. Butterworth (1995). The effects of future consumption by the Cape fur seal on catches and catch rates of the Cape hakes. 4. Modelling the biological interaction between Cape fur seals *Arctocephalus pusillus pusillus* and the Cape hakes *Merluccius capensis* and *M. paradoxis*. *South African Journal of Marine Science*, **16**:255–285.

Punt, A.E. and R. Hilborn (1997). Fisheries stock assessment and decision analysis: the Bayesian approach. *Reviews in Fish Biology and Fisheries*, **7**:35–63.

Punt, A.E. and A.D.M. Smith (2001). The gospel of maximum sustainable yield in fisheries management: birth, crucifixion and reincarnation. In Reynolds, J.D., G.M. Mace, K.R. Redford, and J.R. Robinson (eds.), *Conservation of Exploited Species*. Cambridge, UK: Cambridge University Press, pp. 41–66.

Purcell, J.E.C., W.M.C. Graham, and H.J.C. Dumont (2001). Proceedings of the International Conference on Jellyfish Blooms. *Hydrobiologia*, **451**:333.

Quinn, T.J. and R.B. Deriso (1999). *Quantitative Fish Dynamics*. Oxford, UK: Oxford University Press.

Rabalais, N.N. and R.E. Turner (2006). Oxygen depletion in the Gulf of Mexico adjacent to the Mississippi River. In Neretin, L.N. (ed.), *Past and Present Marine Water Column Anoxia*. NATO Science Series IV: Earth and Environmental Sciences. Dordrecht, the Netherlands: Kluwer, pp. 225–245.

Rabalais, N.N. R.E. Turner, W.J. Wiseman, Jr., and D.F. Boesch (1991). A brief summary of hypoxia on the northern Gulf of Mexico continental shelf: 1985–1988. In Tyson, R.V. and T.H. Pearson (eds.), *Modern and Ancient Continental Shelf Anoxia*. Geological Society Special Publication 58. London: The Geological Society.

Rabalais, N.N., R.E. Turner, D. Justić, Q. Dortch, and W.J. Wiseman, Jr. (1999). *A Characterization of Hypoxia: Topic 1. Report for the Integrated Assessment on Hypoxia in the Gulf of Mexico*. NOAA Coastal Ocean Program Decision Analysis Series 15. Silver Spring, MD: NOAA Coastal Ocean Program. 167 pp.

Rabalais, N.N., R.E. Turner, and W.J. Wiseman, Jr. (2001). Hypoxia in the Gulf of Mexico. *Journal of Environmental Quality*, **30**:320–329.

Rabalais, N.N., R.E. Turner, and D. Scavia (2002a). Beyond science into policy: Gulf of Mexico hypoxia and the Mississippi River. *BioScience*, **52**:129–142.

Rabalais, N.N., R.E. Turner, and W.J. Wiseman, Jr. (2002b). Gulf of Mexico hypoxia, a.k.a., "The Dead Zone". *Annual Review of Ecology and Systematics*, **33**:235–263.

Ragnarsson, S.Á., A. Jaworski, O.A.L. Paramor, *et al.* (2003). *European Fisheries Ecosystem Plan: The North Sea Significant Web*. EU Project number Q5RS-2001-016585. Reykjavik, Iceland: Marine Research Institute.

Restrepo, V.R. (1999). *Proceedings of the Fifth National NMFS Stock Assessment Workshop: Providing Scientific Advice to Implement the Precautionary Approach Under the Magnuson-Stevens Fishery Conservation and Management Act*. NOAA Technical Memorandum NMFS-F/SPO 40, 161 pp.

Restrepo, V.R., G.G. Thompson, P.M. Mace, *et al.* (1998). *Technical Guidance on the Use of Precautionary Approaches to Implementing National Standard 1 of the Magnuson-Stevens Fishery Conservation and Management Act*. NOAA Technical Memorandum NMFS-F/SPO 31, 54 pp.

Rice, J.C. (2000). Evaluating fishery impacts using metrics of community structure. *ICES Journal of Marine Science*, **57**:682–688.

Rice, J.C. (2003). Environmental health indicators. *Ocean and Coastal Management*, **46**:235–259.

Rice, J.C. and M.-J. Rochet (2005). A framework for selecting a suite of indicators for fisheries management. *ICES Journal of Marine Science* **53**:1214–1225.

Richards, R.A. and P.J. Rago (1999). A case history of effective fishery management: Chesapeake Bay striped bass. *North American Journal of Fisheries Management*, **19**:356–375.

Ricker, W.E. (1954). Stock and recruitment. *Journal of the Fisheries Research Board of Canada*, **11**:559–623.

Roberts, C.M. and J.P. Hawkins (1999). Extinction risk in the sea. *TRENDS in Ecology and Evolution*, **14**:241–248.

Rochet, M.-J. and J.C. Rice (2009). Simulation-based management strategy evaluation: ignorance disguised as mathematics? *ICES Journal of Marine Science*, **66**:754–762.

Rochet, M.-J. and V.M. Trenkel (2003). Which community indicators can measure the impact of fishing? *Canadian Journal of Fisheries and Aquatic Sciences*, **60**:86–99.

Rose, K.A. (2000). Why are quantitative relationships between environmental quality and fish populations so elusive? *Ecological Applications*, **10**:367–385.

Rosen, D.A.S. and A.W. Trites (2000). Pollock and the decline of Steller sea lions: testing the junk-food hypothesis. *Canadian Journal of Zoology*, **78**:1243–1250.

Rosenberg, A.A. (2003). Managing to the margins: the overexploitation of fisheries. *Frontiers in Ecology and the Environment*, **1**:102–106.

Rosenberg, A.A. (2007). Fishing for certainty. *Nature*, **449**:989.

Rosenberg, A.A., M.J. Fogarty, M.P. Sissenwine, J.R. Beddington, and J.G. Shepherd (1993). Achieving sustainable use of renewable resources. *Science*, **262**:828–829.

Rosenzweig, M.L. and R.H. MacArthur (1963). Graphical representation and stability conditions of predator-prey interactions. *American Naturalist*, **97**:209–223.

Sainsbury, K.J. (1988). The ecological basis of multispecies fisheries and management of a demersal fishery in tropical Australia. In Gulland, J.A. (ed.), *Fish Population Dynamics*. London: John Wiley, pp. 349–382.

Sainsbury, K.J. (1991). Application of an experimental approach to management of a tropical multispecies fishery with highly uncertain dynamics. *International Council for the Exploration of the Sea Marine Science Symposium*, **193**:301–320.

Sainsbury, K.J. (1998). Living marine resource assessment for the 21st Century: what will be needed and how will it be provided? In Funk, F., I.T.J. Quinn, J. Heifetz, *et al.* (eds.), *Fishery Stock Assessment Models*. Alaska Sea Grant College Program Report AK-SG-98-01. Fairbanks, AK: University of Alaska, pp. 1–40.

Sainsbury, K.J. and U.R. Sumaila (2003). Incorporating ecosystem objectives into management of sustainable marine fisheries, including "best practice" reference points and use of marine protected areas. In Sinclair, M. and G. Valdimarson (eds.), *Responsible Fisheries in the Marine Ecosystem*. Wallingford, UK: CAB International, pp. 343–361.

Sainsbury, K.J., R.A. Campbell, R. Lindholm, and A.W. Whitelaw (1997). Experimental management of an Australian multispecies fishery: examining the possibility of trawl induced habitat modification. In Pikitch, E.K., D.D. Huppert, and M.P. Sissenwine (eds.), *Global Trends: Fisheries Management*. American Fisheries Society Symposium 20. Washington, DC: AFS, pp. 107–112.

Sainsbury, K.J., A.E. Punt, and A.D.M. Smith (2000). Design of operational management strategies for achieving fishery ecosystem objectives. *ICES Journal of Marine Science*, **57**:731–741.

Schaefer, M.B. (1954). Some aspects of the dynamics of populations important to the management of commercial marine fisheries. *Bulletin of the Inter-American Tropical Tuna Commission*, **1**:27–56.

Schaefer, M.B. (1956). The scientific basis for a conservation program. In *Papers presented at the International Technical Conference on the Conservation of the Living Resources of the Sea*. United Nations Publication 1956.II.B.1. New York: United Nations, pp. 15–55.

Schaefer, M.B. (1957). A study of the dynamics of the fishery for yellowfin tuna in the eastern tropical Pacific Ocean. *Bulletin of the Inter-American Tropical Tuna Commission*, **2**:247–285.

Scheffer, M., S. Carpenter, and B. de Young (2005). Cascading effects of overfishing marine systems. *TRENDS in Ecology and Evolution*, **20**:579–581.

Schiller, A., C.T. Hunsaker, M.A. Kane, *et al.* (2001). Communicating ecological indicators to decision makers and the public. *Conservation Ecology*, **5**:1–19.

Schnute, J.T. and R. Haigh (2006). Reference points and management strategies: lessons from quantum mechanics. *ICES Journal of Marine Science*, **63**:4–11.

Schnute, J.T., M.N. Maunder, and J.N. Ianelli (2007). Designing tools to evaluate fishery management strategies: can the scientific community deliver? *ICES Journal of Marine Science*, **64**:1077–1084.

Scott, A. D. (1989). Conceptual origins of rights based fishing. In Neher, P. A., R. Arnason and N. Mollett (eds.), *Rights Based Fishing*. Boston, MA: Kluwer Academic Publishers, pp. 11–38.

Selgeby, J. H., R. L. Eshenroder, C. C. Krueger, J. E. Marsden, and R. L. Pycha (1995). International conference on restoration of lake trout in the Laurentian Great Lakes. *Journal of Great Lakes Research*, **21** (Suppl. 1).

Sette, O. E. (1943). Studies on the Pacific pilchard or sardine (*Sardinops caerulea*). In *Structure of a Research Program to Determine How Fishing Affects the Resources*. Special Scientific Report 19. Washington, DC: U.S. Fish and Wildlife Service.

Shiganova, T. A., Z. A. Mirzoyan, E. A. Studenikina, S. P. Volovick, *et al.* (2001). Population development of the invader ctenophore *Mnemiopsis leidyi*, in the Black Sea and in other seas of the Mediterranean basin. *Marine Biology*, **139**:431–445.

Sibert, J., J. Hampton, P. Kleiber, and M. Maunder (2006). Biomass, size, and trophic status of top predators in the Pacific Ocean. *Science*, **314**:1773–1776.

Sissenwine, M. P. (1978). Is MSY an adequate foundation for optimum yield? *Fisheries*, **3**:22–42.

Sissenwine, M. P. and P. M. Mace (2003). Governance for responsible fisheries: an ecosystem approach. In Sinclair, M. and G. Valdimarsson (eds.), *Responsible Fisheries in the Marine Ecosystem*. Wallingford, UK: CAB International, pp. 363–390.

Sissenwine, M. P. and S. Murawski (2004). Moving beyond "intelligent tinkering": advancing an ecosystem approach to fisheries. In Browman, H. I. and K. I. Stergiou (eds.), Perspectives on ecosystem-based approaches to the management of marine resources. *Marine Ecology Progress Series*, **274**: 291–295.

Sissenwine, M. P. and J. G. Shepherd (1987). An alternate perspective on recruitment overfishing and biological reference points. *Canadian Journal of Fisheries and Aquatic Science*, **44**:913–918.

Skud, B. E. (1975). *Revised Estimates of Halibut Abundance and the Thompson–Burkenroad Debate*. Report of the International North Pacific Halibut Commission, Seattle, WA, 56 pp.

Slocombe, D. S. (1993). Implementing ecosystem-based management. *BioScience*, **43**:612–622.

Smith, A. D. M., K. J. Sainsbury, and R. A. Stevens (1999). Implementing effective fisheries management systems: management strategy evaluation and the Australian partnership approach. *ICES Journal of Marine Science*, **56**:967–979.

Smith, A. D. M., E. J. Fulton, A. J. Hobday, D. C. Smith, and P. Shoulder (2007). Scientific tools to support the practical implementation of ecosystem-based fisheries management. *ICES Journal of Marine Science*, **64**:633–639.

Smith, A. D. M., D. C. Smith, G. N. Tuck, *et al.* (2008). Experience in implementing harvest strategies in Australia's south-eastern fisheries. *Fisheries Research*, **94**:373–379.

Smith, T. D. (1994). *Scaling Fisheries: The Science of Measuring the Effects of Fishing, 1855–1955*. New York: Cambridge University Press.

Smith, T. D. (1998). "Simultaneous and complementary advances": mid-century expectations of the interaction of fisheries science and management. *Reviews in Fish Biology and Fisheries*, **8**:335–348.

Smith, T. D. (2002). The Woods Hole bottom-trawl resource survey: development of fisheries-independent multispecies monitoring. In Anderson, E. D. (ed.), *100 Years of Science under ICES: Papers from a Symposium Held in Helsinki*. ICES Marine Science Symposia 215. Copenhagen, Denmark: ICES, pp. 474–482.

Smith, T. D. and J. S. Link (2005). Autopsy your dead...and living: a proposal for fisheries science, fisheries management and fisheries. *Fish and Fisheries*, **6**:73–87.

Solow, R. (1991). Sustainability: an economist's perspective. In Dorfman, R. and N. Dorfman (eds.), *Economics of the Environment*. New York: W. W. Norton.

Steele, J.H. (1985). A comparison of terrestrial and marine ecological systems. *Nature*, **313**:355–358.

Steele, J.H. (1991). Can ecological theory cross the land-sea boundary? *Journal of Theoretical Biology*, **153**:425–436.

Stefánsson, G. and Ó.K. Pálsson (1998). A framework for multispecies modelling of Boreal systems. *Reviews in Fish Biology and Fisheries*, **8**:101–104.

Stenson, G.B. and M.O. Hammill (2006). *Quantifying Uncertainty in Estimates of Atlantic Cod* (Gadus morhua) *Consumption by Harp Seals* (Phoca groenlandica). Canadian Science Advisory Secretariat (CSAS) Research Document 2004/089. Ottawa, Canada: DFO, CSAS, 32 pp.

Stenson, G.B. and E.A. Perry (2001). *Incorporating Uncertainty into Estimates of Atlantic Cod* (Gadus morhua), *Capelin* (Mallotus villosus) *and Arctic Cod* (Boreogadus saida) *Consumption by Harp Seals* (Pagophilus groenlandicus) *in NAFO Divisions 2J3KL*. Canadian Science Advisory Secretariat (CSAS) Research Document, 2001/074. Ottawa, Canada: DFO, CSAS, pp. 1–34.

Stenson, G.B., M.O. Hammill, and J.W. Lawson (1997). Predation by harp seals in Atlantic Canada: preliminary consumption estimates for Arctic cod, capelin and Atlantic cod. *Journal of Northwest Atlantic Fishery Science*, **22**:137–154.

Stobutzki, I.C., M.J. Miller, P. Jones, and J.P. Salini (2001a). Bycatch diversity and variation in a tropical Australian penaeid fishery: the implications for monitoring. *Fisheries Research*, **53**:283–301.

Stobutzki, I.C., M.J. Miller, and D. Brewer (2001b). Sustainability of fishery bycatch: a process for assessing highly diverse and numerous bycatch. *Environmental Conservation*, **28**:167–181.

Stobutzki, I.C., M.J. Miller, D.S. Heales, and D.T. Brewer (2002). Sustainability of elasmobranchs caught as bycatch in a tropical prawn (shrimp) trawl fishery. *Fishery Bulletin*, **100**:800–821.

Suter, G.W. (1993). *Ecological Risk Assessment*. Boca Raton, FL: Lewis Publishers.

Symes, D. and J. Phillipson (2009). Whatever became of social objectives in fisheries policy? *Fisheries Research*, **95**:1–5.

Tasker, M.L., C.J. Camphuysen, J. Cooper, *et al.* (2000). The impacts of fishing on marine birds. *ICES Journal of Marine Science*, **57**:531–547.

Thomas, J.W. (2000). From managing a deer herd to moving a mountain: one pilgrim's progress. *Journal of Wildlife Management*, **64**:1–10.

Thomas, J.W. (2004). *The Journals of a Forest Service Chief*. University of Washington Press.

Thompson, W.F. (1922). The marine fisheries, the state and the biologist. *Scientific Monthly*, **15**:542–550.

Tjelmeland, S. and B. Bogstad (1998). MULTSPEC: a review of a multispecies modelling project for the Barents Sea. *Fisheries Research*, **37**:127–142.

Townsend, H.M., J.S. Link, K.E. Osgood, *et al.* (2008). *Report of the NEMoW (National Ecosystem Modeling Workshop)*. NOAA Technical Memorandum NMFS-F/SPO-87, 93 pp.

Trenkel, V.M. and M.-J. Rochet (2003). Performance of indicators derived from abundance estimates for detecting the impact of fishing on a community. *Canadian Journal of Fisheries and Aquatic Sciences*, **60**:67–85.

Trites, A.W., P.A. Livingston, S. Mackinson, *et al.* (1999). *Ecosystem Change and the Decline of Marine Mammals in the Eastern Bering Sea: Testing the Ecosystem Shift and Commercial Whaling Hypotheses*. Fisheries Centre Research Reports 7(1). Vancouver, Canada: University of British Columbia, 106 pp.

Tsou, T.S. and J. Collie (2001a). Predation-mediated recruitment in the Georges Bank fish community. *ICES Journal of Marine Science*, **58**:994–1001.

Tsou, T.S. and J. Collie (2001b). Estimating predation mortality in the Georges Bank fish community. *Canadian Journal of Fisheries and Aquatic Science*, **58**: 908–922.

Tudela, S., M. Coll, and I. Palomera (2005). Developing an operational reference framework for fisheries management on the basis of a two-dimensional index of ecosystem impact. *ICES Journal of Marine Science*, **62**:585–591.

Tyrrell, M.C., J.S. Link, H. Moustahfid, and W.J. Overholtz (2009). Evaluating the effect of predation mortality on forage species population dynamics in the Northwest Atlantic continental shelf ecosystem: an application using multispecies virtual population analysis. *ICES Journal of Marine Science*, **65**:1689–1700.

Tyrrell, M.C., J.S. Link, and H. Moustahfid (2010). The importance of including predation in some fish population models: implications for biological reference points. *Fisheries*, in press.

Turner, R.E. and N.N. Rabalais (1994). Coastal eutrophication near the Mississippi river delta. *Nature*, **368**:619–621.

Ulanowicz, R.E. (1986). *Growth and Development: Ecosystem Phenomenology*. New York: Springer.

Ulanowicz, R.E. (1997). *Ecology: The Ascendant Perspective*. New York: Columbia University Press.

Ulanowicz, R.E. (2004). Quantitative methods for ecological network analysis. *Computational Biology and Chemistry*, **28**:321–339.

Ulanowicz, R.E. and J.J. Kay (1991). A package for the analysis of ecosystem flow networks. *Environmental Modelling and Software*, **6**:131–142.

Ulanowicz, R.E. and J.S. Norden (1990). Symmetrical overhead in flow and networks. *International Journal of Systems Science*, **21**:429–437.

Ulanowicz, R.E. and C.J. Puccia (1990). Mixed trophic impacts in ecosystems. *Coenoses*, **5**:7–16.

USCOP (U.S. Commission on Ocean Policy) (2004). *An Ocean Blueprint for the 21st Century*. Washington, DC: U.S. Commission on Ocean Policy.

USDA (U.S. Department of Agriculture) (2009). *Census of Agriculture 2007*. Geographic Area Series, Part 51, Vol. 1 AC-07-A-51. Washington, DC: USDA, 739 pp.

Utne, I.B. (2006). Systems engineering principles in fisheries management. *Marine Policy*, **30**:624–634.

Vézina, A.F. and M. Pahlow (2003). Reconstruction of ecosystem flows using inverse methods: how well do they work? *Journal of Marine Systems*, **40–41**:55–77.

Viaroli, P., P. Lasserre, and P. Campostrini (eds.) (2007). *Lagoons and Coastal Wetlands in the Global Change Context: Impacts and Management Issues*. Dordrecht, The Netherlands: Springer.

Vinther, M. (2001). Ad hoc multispecies tuning applied for the Baltic and the North Sea fish stocks. *ICES Journal of Marine Science*, **58**:311–320.

Volterra, V. (1926). Fluctuations in the abundance of a species considered mathematically. *Nature*, **118**:558–560.

Walters, C. (1986). *Adaptive Management of Renewable Resources*. Caldwell, NJ: The Blackburn Press.

Walters, C.J. and J.S. Collie (1988). Is research on environmental factors useful to fisheries management? *Canadian Journal of Fisheries and Aquatic Science* **45**:1848–1854.

Walters, C.J. and J.F. Kitchell (2001). Cultivation/depensation effects on juvenile survival and recruitment: implications for the theory of fishing. *Canadian Journal of Fisheries and Aquatic Science*, **58**:39–50.

Walters, C.J., V. Christensen, and D. Pauly (1997). Structuring dynamic models of exploited ecosystems from trophic mass-balance assessments. *Reviews in Fish Biology and Fisheries*, **7**:139–172.

Walters, C.J., V. Christensen, S.J. Martell, and J.F. Kitchell (2005). Possible ecosystem impacts of applying MSY policies from single-species assessment. *ICES Journal of Marine Science*, **62**:558–568.

Ward, T., D. Tarte, E. Hegerl, and K. Short (2002). *Ecosystem-based Management of Marine Capture Fisheries*. Sydney, Australia: World Wide Fund for Nature.

Ware, D.M. and R.E. Thomson (2005). Bottom-up ecosystem trophic dynamics determine fish production in the Northeast Pacific. *Science*, **308**:1280–1284.

Waskow, A. (1995). *Down to Earth Judaism*. New York: Morrow.

WCED (United Nations World Commission on Environment and Development). (1987). *Our Common Future: Report of the World Commission on Environment and Development* (a.k.a. Brundtland Report). Published as Annex to General Assembly Document A/42/427, Development and International Co-operation: Environment. Oxford, UK: Oxford University Press.

Whipple, S., J.S. Link, L.P. Garrison, and M.J. Fogarty (2000). Models of predation and fishing mortality in aquatic ecosystems. *Fish and Fisheries*, **1**:22–40.

Wicklum, D. and R.W. Davies (1995). Ecosystem health and integrity? *Canadian Journal of Botany*, **73**:997–1000.

Witherell, D., C. Pautzke, and D. Fluharty (2000). An ecosystem-based approach for Alaska groundfish fisheries. *ICES Journal of Marine Science*, **57**:771–777.

Worm, B. and R.A. Myers (2003). Meta-analysis of cod-shrimp interactions reveals top-down control in oceanic food webs. *Ecology*, **84**:162–173.

Worm, B., M. Sandow, A. Oschlies, H.K. Lotze, and R.A. Myers (2005). Global patterns of predator diversity in the open oceans. *Science*, **309**:1365–1369.

Yaffee, S.L. (1996). Ecosystem management in practice: the importance of human institutions. *Ecological Applications*, **6**:724–727.

Yaffee, S.L. (1999). Three faces of ecosystem management. *Conservation Biology*, **13**:713–725.

Yaffee, S.L., A.F. Phillips, I.C. Frentz, *et al.* (1996). *Ecosystem Management in the United States: An Assessment of Current Experience*. Washington, DC: Island Press.

Yodzis, P. (1994). Predator-prey theory and management of multispecies fisheries. *Ecological Applications*, **4**:51–58.

Yodzis, P. (2000). Diffuse effects in food webs. *Ecology*, **81**:261–266.

Yodzis, P. (2001). Must top predators be culled for the sake of fisheries? *TRENDS in Ecology and Evolution*, **16**:78–84.

Zhou, S. and S.P. Griffiths (2008). Sustainability assessments for fishing effects (SAFE): a new quantitative ecological risk assessment method and its application to elasmobranch bycatch in an Australian trawl fishery. *Fisheries Research*, **91**:56–68.

Index